REPAIRING
FRENCH PENDULUM CLOCKS

STEVEN G. CONOVER

CLOCKMAKERS NEWSLETTER, READING

Dedicated to my wife, Karen

ii

Published by Clockmakers Newsletter
203 John Glenn Avenue
Reading, PA 19607

Text, photos and drawings Copyright 2006 by Steven G. Conover

No part of this book may be reproduced in any manner without written permission from the publisher. Printed in the USA.

ISBN 978-0-9752574-1-8

CONTENTS

INTRODUCTION

Clock repairers do not often encounter time-keepers that are better made than French pendulum clocks. The most common example is the familiar round movement called *pendule de Paris.* The overall quality and finish are such that the clock's worst enemies are careless, untrained repairers, not the effects of time or use. I cannot think of a style of old clock that will run for a longer stretch of time, displaying less wear, than the French clock. The worst damage that one usually sees is gearing damaged by a mainspring that has shattered, releasing all its power at once. Otherwise, most French clocks will run for years without even being oiled. Eventually, lack of care takes its toll on the movement in the form of worn pivots and pivot holes.

Perhaps a French clock should not be the first or second one you disassemble and repair. As a new beginner many years ago, I avoided French clocks for the first three years. But I hope that others, armed with the information in this book, will work on French clocks much sooner than that. I think the main worry is that of bending or breaking a fine pivot during the assembly procedure. With a little practice, this fear eases. Work at a comfortable bench with good light, use tweezers to move the arbors, and be patient. Clocks which are as well made as these are not as difficult to put together as one might think. In addition, the removable cocks on both the time and strike trains ease the task of assembly and adjustment.

My approach in this, my ninth book, is to show many sequential photos of a process such as reassembly, rather than to present a single, large drawing of the parts. I have also taken the step of eliminating figure numbers for the illustrations, hoping to achieve a freer flowing presentation. Finally, I have continued my practice of showing specific examples, rather than just general types. As always, the examples were carefully selected.

There are several repairs that I have chosen not to include in this book. The replacement of small French pivots is better left to the specialist or the most advanced repairer. In the same way, many clock repair shops will want to "farm out" wheel and pinion cutting and the retoothing of barrel wheels. This is not to say that these skills should not be learned. It is just that most shops cannot devote the time to accomplish these tasks, nor can they afford to own the necessary equipment.

French clocks are varied and exciting to collect, and I am sure they will remain a staple in the repairer's workshop.

Steven G. Conover
Reading, Pennsylvania, August 2006

SOURCES

de Carle, Donald. *Watch & Clock Encyclopedia.* New York: Bonanza Books, 1977.

Penman, Laurie. *Practical Clock Escapements.* Ashbourne, Derbyshire, England: Mayfield Books, 1998.

Seymour, Lawrence A. "Morbier Clocks." Bulletin of the N.A.W.C.C., Inc. 157 (April 1972): 270-290.

Seymour, Lawrence A. "Morbier Clocks." Bulletin of the N.A.W.C.C., Inc. 162 (February 1973): 962-976.

Seymour, Lawrence A. "Morbier Clocks." Bulletin of the N.A.W.C.C., Inc. 172 (October 1974): 566-589.

Thorpe, Nicolas M. *The French Marble Clock.* Colchester, Essex, England: N.A.G. Press, 1990.

ACKNOWLEDGMENTS

I would like to express my thanks to Paul Corn for his advice so freely given, especially on the silk thread suspension; to Lou Orsini, for the loan of the clock featured in Chapter 6; and to an owner I will call Edward M. for the use of the clock featured in Chapter 3—I do not wish to reveal the identity of a private collector.

French timepiece with black slate case and grayish marble accent pieces, 8-1/2" tall.

A BLACK MARBLE TIMEPIECE

The best French clock to start on is a timepiece, that is, a clock which does not strike. There are only half the parts to deal with, and the repair will probably be much easier as a result.

The clock shown at the top of the page was a clock mart purchase in 1993. I was attracted to the clock by its low price, combined with the nice black case that had few defects. In fact, it was free of the severe chipping and damage that is common in such clocks. Beyond that, a careful cleaning and polishing of the case revealed gray marble edge strips on the sides and base of the clock. These had been completely obscured by dirt; they appeared black! I have rarely found greater pleasure in cleaning and polishing a case.

A bad dial is the downfall of many an otherwise good French clock. This clock's enamel dial was in good condition, except for slight chipping around the keyhole; there were none of the spidery cracks or chipped-away sections of enamel that are to be expected in a century-old dial. I installed a brass dial grommet in the keyhole and found that it greatly improved the appearance.

I ran this clock for about half of the 13 intervening years between my original repair and the present day. In this chapter, I will review all the repairs that were undertaken. The current tasks included another cleaning, of course. In addition, the second arbor's pivots had become worn, and I had to stone and polish them. Bushings were installed in the corresponding holes.

A view through the open back door.

This view shows the movement as it appeared recently, upon removal for an overdue cleaning.

To remove the hands, begin by pulling out the taper pin from the center arbor. Pulling will work better than squeezing out the very fine taper pin typically used in a French clock center arbor.

Separate the movement from the dial by removing three taper pins.

One of the first steps in any repair is to let down the mainspring(s). In this clock, there is just one mainspring. The photo shows that I used a toothpick—which was about to break—to pry the click away from the clickwheel. A screwdriver can be used, but be careful to avoid scratching the parts. A letdown key is seen at the right side of the photo. These are available from parts suppliers; don't try to let down a mainspring without one! It isn't safe.

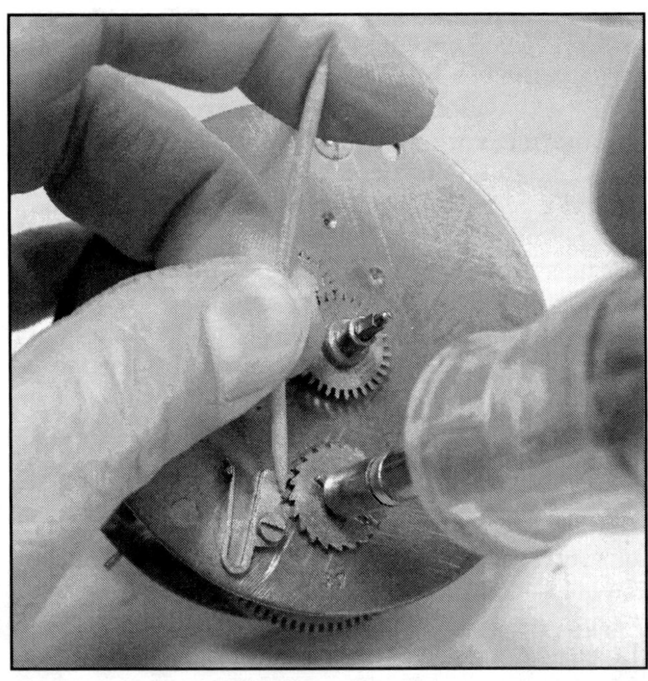

A single screw holds the back cock onto the back plate. Once the screw is out, take off the back cock to expose the pallet arbor; remove the pallet unit and set it safely aside.

Remove the motion work parts from the front of thc movement. Two of the parts are a bit out of the ordinary in this movement. The click and clickspring are made from a single piece of steel; the clickwheel is retained by a taper pin, rather than a separate brass cover with a set screw.

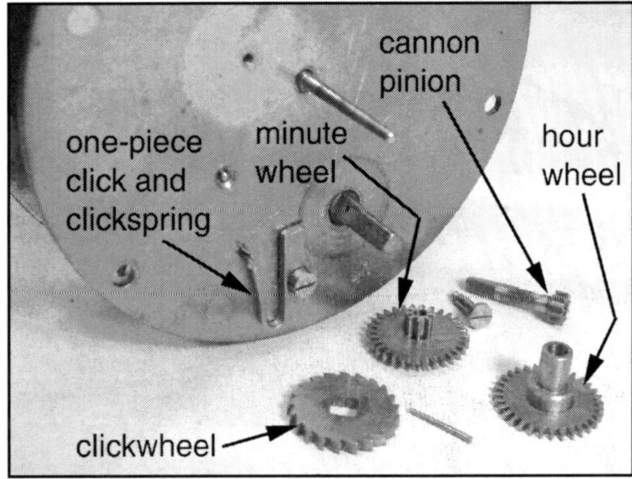

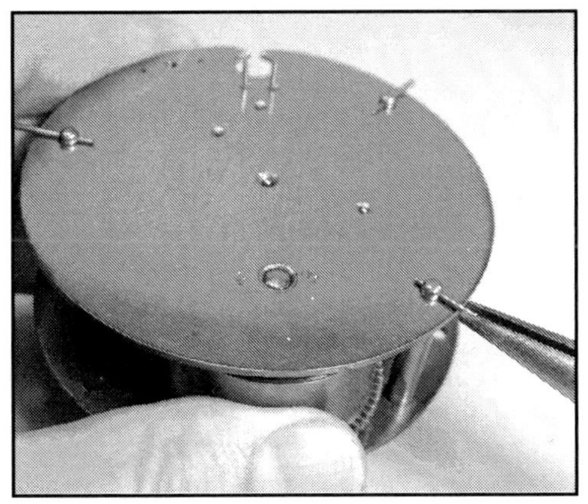

Pull or squeeze out the pillar pins with pliers. Bear in mind that the pins may have been driven in with excessive force. Try not to break off a pin. It is a tiresome job to file away the parts of the pin extending from the pillar. You will have to drive out or drill out the part that remains in the hole in the pillar. Luckily, I had no such trouble when I disassembled this movement. Always use steel taper pins, not brass.

Disassembly reveals relatively few parts in this simple movement.

Bushings

When I first repaired the movement, I found that all of the pivot holes, including those of the winding arbors, had been bushed. So it appeared the movement was either badly worn or it was unnecessarily bushed. Even so, some of the pivots fitted loosely in the KWM-size bushings. This meant the bushings were either worn from use or badly fitted in the first place. No attempt had been made to finish off the oil sinks to blend in with the plates. On page 21, I show my method for finishing the oil sinks in press-fit bushings for a pleasing appearance.

In addition, certain bushings were not flush with the inside surface of the plate. They were either recessed or extended from the inside

surface of the plate. This was an attempt to ease a gearing problem. It is an approach that has its place, I suppose, but I prefer to correct a gearing problem, rather than just hide it.

Remounting a Wheel

If there was one major problem in this movement, it was the way the third wheel meshed with the escape pinion. The drawing shows how the third wheel **b** contacted the escape pinion **a** at its very end. The two still meshed and the clock would run, but the escape pinion was badly worn down in the contact area. Fortunately, the third wheel teeth still looked unworn.

There had to be a reason for this poor alignment of the wheel and pinion. It is possible that the original third wheel, pinion, and arbor unit was lost or damaged and replaced with another unit with the same tooth counts. The replacement wheel did not line up well with the escape pinion. This theory is supported by the fact that the third wheel also ran very close to the center pinion.

The best way to correct the problem was to move the third wheel further back on its arbor. This would allow it to mesh toward the center portion of the escape pinion, in an unworn area.

The wheel was attached to the end of the pinion, with the leaves pressed into the center hole of the wheel. The overall plan was to remove the wheel from the pinion and install a custom made brass collet in its place. The wheel, in turn, would be mounted to the new collet. This would move the wheel a certain distance along the arbor, and it would mesh with the escape pinion at the middle portion, rather than at the end.

Here is how it was done. A length of brass rod (aluminum would do as well) was gripped in the Sherline 3-jaw chuck. The end of the rod was faced off. Next, a hole was drilled in the rod, slightly larger than the expected diameter of the hole that would be bored in the center of the wheel. A boring tool was then used to make a recess in the end of the rod, which would hold the wheel. The recess was carefully enlarged until the wheel snapped in place. Shellac of the type used to cement pallet jewels, or Super Glue, can be used to hold the wheel if necessary. However, the fit must be close in order to maintain the concentricity of the center hole with the O.D. of the wheel.

The next step was to bore out the center hole of the wheel, because the existing hole was deeply cut and distorted from being staked onto the end of the pinion. The photo shows this setup. The boring tool was set on the center height of the lathe. Light cuts were taken to

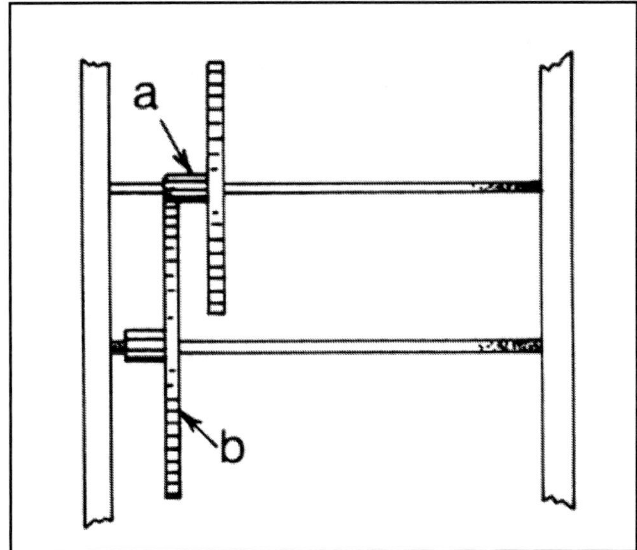

enlarge the hole until it was clean and sharp. When this was complete, the wheel was removed from the recess in the brass rod, and the rod was taken from the chuck.

The new wheel collet was made from a piece of small diameter brass rod. The basic design is shown in the sketch. The larger diameter of the wheel collet had to be greater than that of the pinion, because the collet was to be pushed up against it. The collet's smaller diameter was 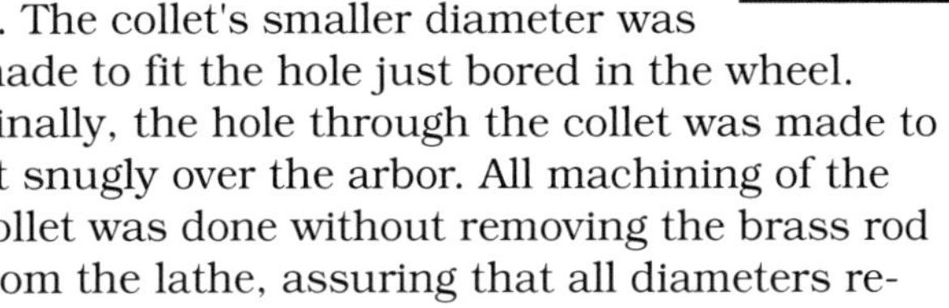 made to fit the hole just bored in the wheel. Finally, the hole through the collet was made to fit snugly over the arbor. All machining of the collet was done without removing the brass rod from the lathe, assuring that all diameters remained concentric.

...continued on page 6

Now the wheel was pressed onto the collet, although it was not riveted as yet. The mounted wheel was then placed on the arbor. It could be seen that the wheel collet had to be counterbored so that it would fit over the smaller section of the pinion where the wheel had originally been riveted. The collet would now push flush against the finished diameter of the pinion. In the process, the position of the wheel became just right, so that the third wheel would mesh in the middle of the escape pinion.

The remounting of the wheel was nearly finished. All that remained was to fasten the parts together. I staked the wheel collet onto the arbor; it became tight just as it slipped over the end of the pinion and pressed against the end of the finished leaves. Finally, the wheel was lightly riveted onto the collet.

The revised gearing was tried in the movement, and the result is shown in the photo. I have also pointed out the place where the wheel meshed with the pinion before the wheel was remounted. This repair is an example of basic lathe work that can produce a professional result.

A previous repairer had tried to make the incorrect gearing work by placing the bushings for the third arbor in a certain way. The front bushing extended inward, and the rear bushing was recessed in the back plate. That approach did not look professional, and it made the clock more difficult to assemble. It also came far short of achieving the desired result.

I was happy with the repair method I had devised.

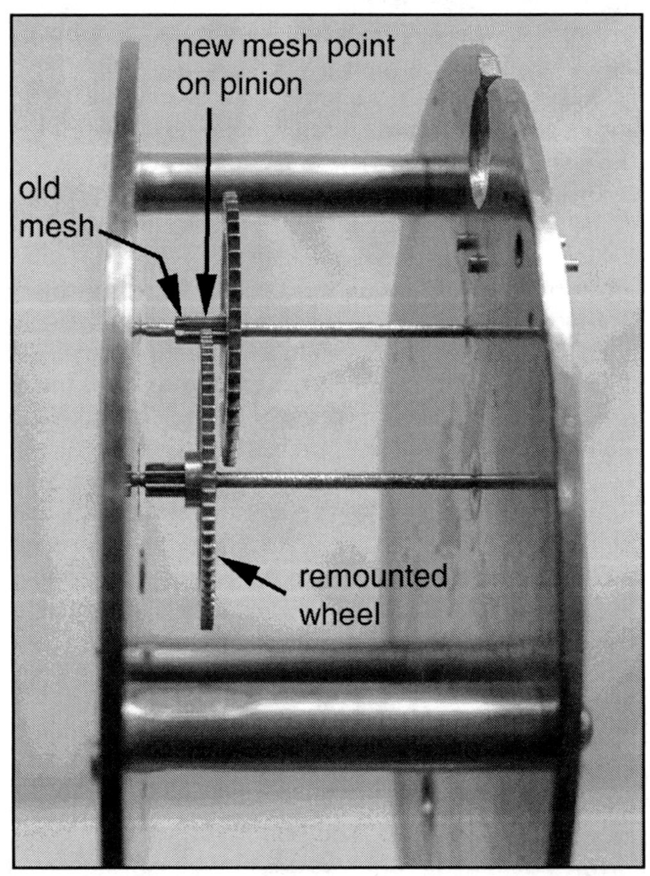

The Mainspring

After I finished the repair in '93, I found that this French timepiece had a poor timekeeping rate. A few months later, I disassembled the movement again so that I could check the mainspring as a possible cause. Specifically, I wanted to know if the old spring (left) had become "set" or weak. Compare it to the much larger uncoiled diameter of a new spring I purchased (right).

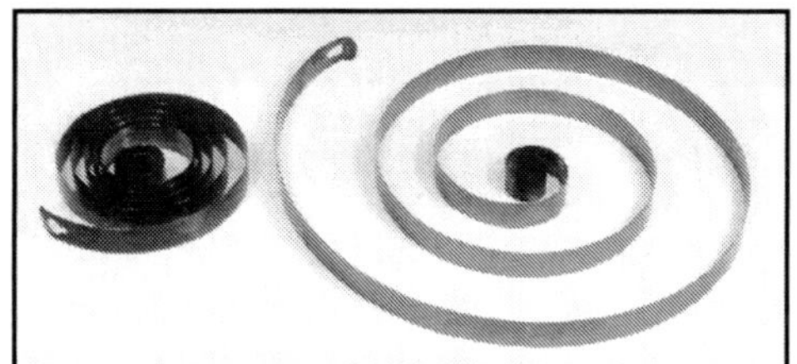

The original spring dimensions were .012" thick x 19/32" wide x 48" long. I checked this length with a formula which will tell us the maximum length spring, of a given thickness, that a particular barrel will hold. A spring that is too long will reduce the number of turns a barrel will make, just as surely as a spring that is too short.

The formula confirmed that the spring in the barrel was the correct length. In applying the formula, a result that is within an inch or so either way, can be called correct.

Mainspring Length in Five Steps

1. Inside diameter of barrel, squared, times .7854

2. Diameter of winding arbor, squared, times .7854

3. Subtract Step #2 from Step #1

4. Divide by 2

5. Divide by the mainspring thickness

The formula works with either metric or English measurements, although you must obviously be consistent in using one system throughout the calculation.

In everyday repairs, the formula will help you to select the best spring. It is often impossible to find the exact new spring to replace a broken one. If it is necessary to change the thickness by .001" either way, use the formula to determine whether the available compromise spring is long enough for the barrel. If the best spring for width and thickness is much too long, the formula will tell you how much of the end of the spring to cut off.

Our featured clock provides an example, as shown on the next page.

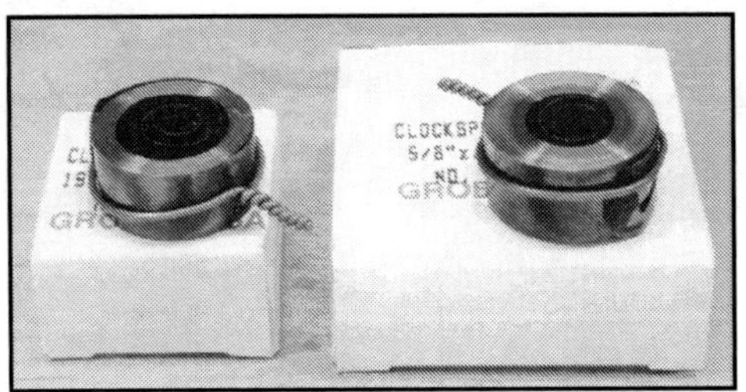

An experiment was conducted on two different mainsprings for this clock.

The first spring I wanted to try was an exact replacement for the old one. The closest spring I could buy had the same thickness and width, but it was 53" long. Having already used the formula on page 7 to double check the ideal length of the original .012" thick spring as 48", I shortened the new one to equal it.

The second spring was .011" thick x 19/32" x 49" long. The formula returned a length of 52" for this thinner spring. For my purpose, I did not worry about the 3" shortfall on the length. I wanted to see if a thinner spring would make the clock run closer to correct time.

First, I found that each spring could rotate the barrel about 7 turns. The thinner spring would probably have made part of another turn, except for the fact that it was three inches short of its ideal length. I calculated that the barrel would turn 5-1/2 times for a 7-day run, or 6-1/2 turns for 8 days.

I had thought that the thinner spring might make this clock's rate more constant, but this was not the case. The graph shows the clock's performance with the new .011" thick spring. It worked about the same with the new .012" spring.

How to Shorten a Mainspring

1. Using a torch, soften the spring, starting at the end and going about 1-1/2" beyond the line you will cut. Do not heat to red; lower heat will do. In the case of shortening to remove a torn hole end, you should cut off about two inches.

2. Cut with shears or a saw.

3. Using the cut-off end as a pattern, use a permanent marker to trace the shape of the hole onto the new end.

4. Use a metal punch to make a new hole in the spring. An alternative is to center punch the hole and drill it, with the spring held in the vise, backed up by a wood block.

5. Use small files to shape the hole the same as the old one. The rule is: do not leave any sharp corners in your shaped hole, or the spring will soon tear there.

6. File away any sharp edges and place the usual bend in the end of the spring to catch the barrel rivet.

The graph shows that the rate, in beats per hour, declined all week. The sharpest decline occurred on the 8th day. When regulated "by hand" instead of with a timer, a clock like this will gain the first half of the week and lose same amount in the second half. This is not desirable, but it is typical of many clocks.

A more recent check with the timer, following the latest cleaning of the movement, shows that the rate wanders considerably from minute to minute. Whatever the reason, this is a patient with an irregular "ticker". A new mainspring was not the answer to its ailment. Still, I think that with careful regulation of the pendulum, the clock's overall weekly performance would be acceptable to most owners.

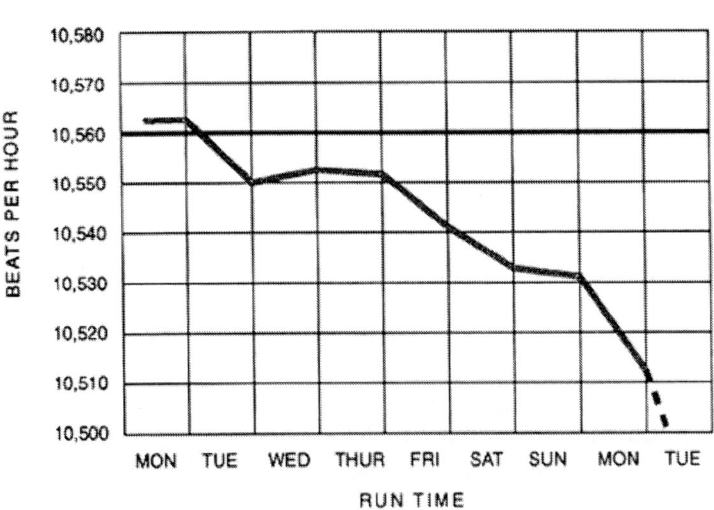

The Escapement

The recoil escapement of our timepiece movement was operable, but there was a deep groove in each of the pallet faces, **d** and **e**. It is not possible to smooth away severe grooves without ruining the pallets. The original surface would be shaved down until it reached the lowest level of the groove. Little benefit would be gained, except that a tooth would no longer have to climb out of the end of a groove. The "drop" of a tooth away from a pallet would still be excessive. Attempting to "close" the solid pallet body **c** in a vise will not work, either. Bending can be successfully employed with bent steel strip pallets, despite some small changes in the pallet angles. But bending the solid French pallet unit is too risky; breakage is very likely.

The best approach for this escapement was to move the thick pallet body further along the pallet arbor to expose a fresh area on each pallet. This is a good way to renew the escapement for years of additional service.

The squared end portion of the pallet arbor **f** was tapered, so I decided to remove the pallet body and then file the tapered arbor very slightly. This would allow the pallet body to move further along the arbor before it tightened against the taper.

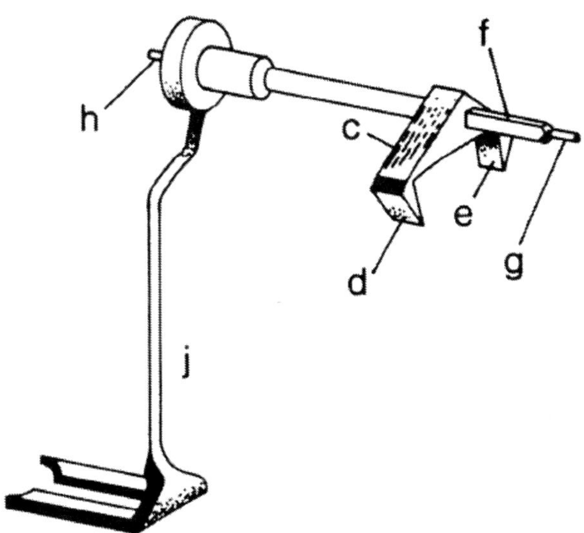

A typical French recoil pallet assembly: **c** pallet body; **d** entry pallet; **e** exit pallet; **f** square portion of pallet arbor; **g** front pivot; **h** rear pivot; **j** crutch.

Removing the pallet from the arbor was the first task. The entire pallet unit was placed in the split stake, as shown in the photo. I used the smallest hole that would allow the round portion of the pallet arbor to pass through. Ron Pfleger, who advised me on this repair, was concerned about using a hand held stake over the front pivot **g** which was only .016" in diameter. With any misalignment of the stake, the shoulder of the pivot could be damaged or the pivot snapped right off.

Accordingly, I used the lathe to make a stake out of brass rod. I drilled the center hole #77, which is .018", to safely clear the pivot.

The drill press, switched off, of course, was used to grip the stake and push straight downward on the pivot shoulder. The pallet body was easy to remove this way.

Next, I carefully filed a small amount of material off the tapered pallet arbor **f**. Then I staked the pallet body back onto the arbor. It moved further along, aligning fresh pallet surfaces with the escape wheel teeth.

Reassembly

The movement consists of far fewer parts than a striking French movement. Support the front plate on a small box, or on a hollow cardboard core as shown in the large photo. This allows the center arbor to project downward without contacting the bench top. Place the arbors into the front plate.

Add the rear plate and begin to fit the rear pivots into their holes. Use a light touch, pushing each pivot into place with tweezers. Any forcing of the pivots may result in a bend or break, but if you are careful, there is no reason for concern.

When the plates are together, install new steel taper pins in the back plate.

Turn to the front of the movement and install the motion work (large photo at right). The parts of the motion work are also shown in the disassembly photo on page 3.

Take the opportunity to oil the front pivot of the center arbor, before adding the cannon pinion. In a similar way, oil the front barrel arbor pivot (inset photo) before installing the clickwheel.

The Regulator

The regulator is worth a look, as it is not a common style. Stamped "Thieble", it works by raising and lowering the top block of the suspension spring. The spring is lightly gripped by the lower part of the brass blocks that make up the body of the regulator. This means that the effective length of the spring is determined by how much of it extends below the blocks. If it is necessary to replace a suspension spring in this type of regulator, the top part of it will have to be carefully fitted.

During the cleaning of the movement, remove the regulator from the back cock; it is held by just one small screw. The regulator can be cleaned assembled and blown dry with compressed air, or it can be further diassembled for cleaning. In this case I took the first approach.

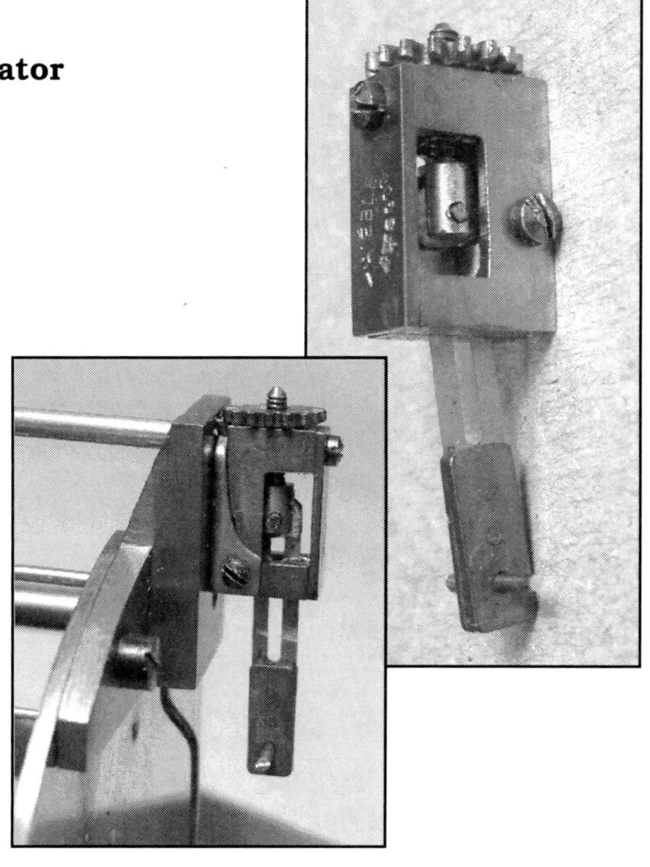

Bluing the Hands

The bluing of clock hands with heat is covered in detail in *Clock Repair Skills*. To summarize here, the hands are first cleaned of all rust and given a high polish. This takes considerable time. Any defect left on the steel will show as a different color or tone when the bluing is done. Wipe the hands with alcohol and do not touch them with your fingers after that. Place one hand in a bluing pan, or its equivalent, resting on a bed of clean brass chips swept from the lathe bench. Some prefer clean sand.

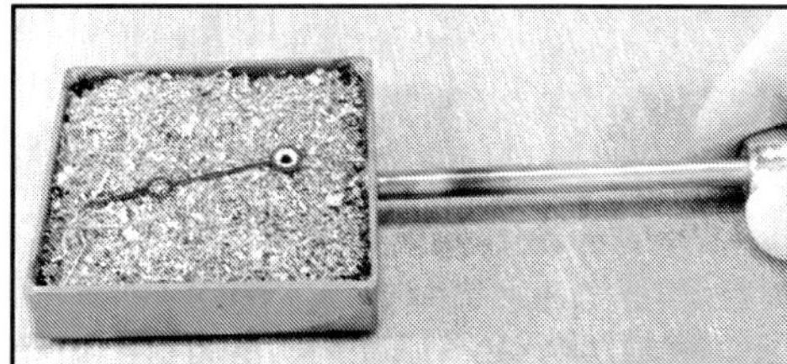

Hold the pan over a propane torch and move it back and forth over the flame for several minutes. The hand will change color, first to pale yellow, moving through straw, brown, and purple. The hand is now close to achieving the deep blue that is desired. When the blue appears over the entire hand, remove it with old tweezers and drop it into water or place it on a piece of glass to cool. If the hand passes to a pale blue before the color changes have been stopped, the entire job must be done over, beginning with the polishing. Coat the finished, cooled hand with oil and wipe with a clean cloth. This enhances the color and prevents rusting.

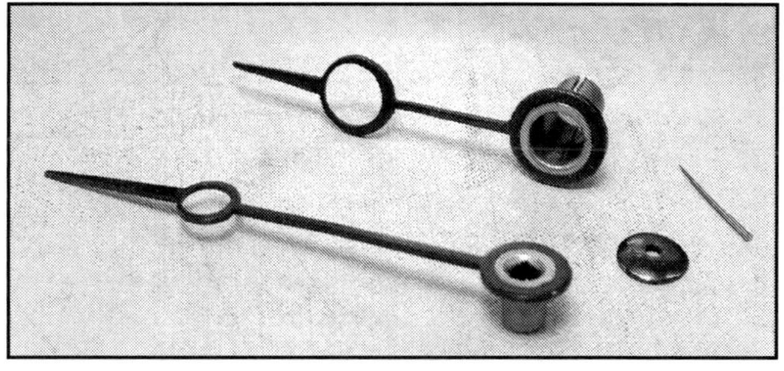

The Movement Straps

Two brass straps attach to the sides of the dial assembly and extend to the rear of the case to hold the movement and dial unit firmly in place. Almost all round French movements use this arrangement. Screws through the rear case bezel (door unit) thread into the straps to exert pressure on the case. The pressure must be firm enough to prevent the entire movement and dial from turning as the movement is wound.

When a strap breaks or the threads strip out, the ingenuity of the clock repairer is brought to bear on the problem. Hex nuts or bits of threaded brass are soldered or glued onto the old straps. These repairs run the gamut from professional to embarrassing.

The solution that had been used on this clock was an inventive shortcut. The repairer took the trouble to replace the straps, but the ones he

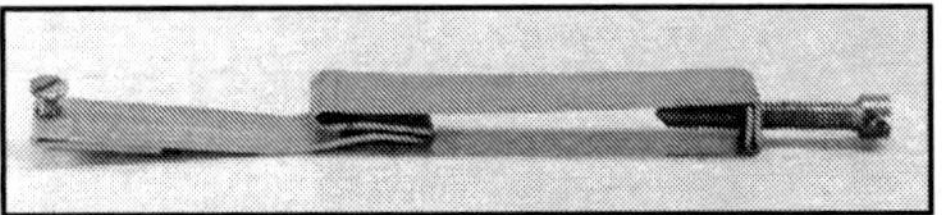

found were too long. So he simply folded them to a length that would fit the case. When the screws were turned, the straps tended to stretch out again, instead of tightening!

New straps shortened to the correct length are a better answer. It is easier to purchase replacement straps and fit them than it is to make them from scratch. The replacements already have the right angle bends and are fitted with screws.

Begin by fitting the movement/dial unit to the case. Set the strap length by allowing sufficient room for the screws to tighten and hold. A good estimate is that the right angle bend in a strap should be about 1/2" from the inside back of the case. Cut the straps.

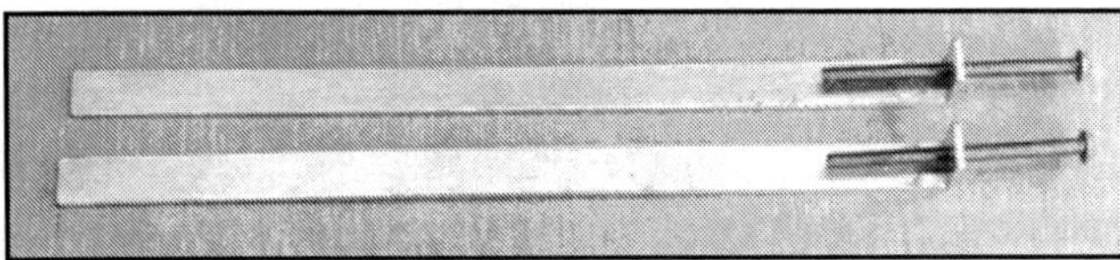

Second, attach the ends of the straps to the dial unit. Many of these are riveted, but ours were screwed. I maintained the same setup by drilling and threading holes for new #2-56 screws.

Whether rivets or screws are used, the attachment should fit freely enough to allow the straps to swivel. This will allow them to pull evenly in the case.

The new straps are shown in the photo at the right, indicated by arrows.

Setting the Beat of the Clock

Setting up this French movement to run is typical of the round French mantel clocks. Start by placing the clock on a level shelf. Hang the pendulum and start it swinging. A small arc will allow you to better assess the beat.

If the ticking sounds even, like that of a metronome, the clock is ready to run.

If the clock is slightly out of beat, loosen the strap screws at the back of the case and twist the entire unit in the case. You will realize that a even a small twist will make the clock go completely in or out of beat. Tighten the screws when the beat sounds correct.

A clock that is badly out of beat cannot be corrected only by twisting the movement/dial unit. It will be necessary to bend the crutch, which is part **j** on the drawing, page 9. To do this safely, let down the time mainspring and remove the back cock and the pallet unit. Now use two pairs of pliers, one to hold and the other to bend the crutch.

Reassemble and place the movement/dial unit back in the case. If the clock is close to being in correct beat, repeat the small twisting adjustment described above to fine tune it. When the work is done, the dial should look straight in the case.

Some French crutches are made to slip on screw threads, so that with some applied overswing they set the beat themselves, making all the foregoing adjustments unnecesary. If you find one of these bound with solder, clean it off to allow it to slip properly again. Other crutches are not self-setting, but the crutch collet allows light finger pressure to move the crutch on the arbor.

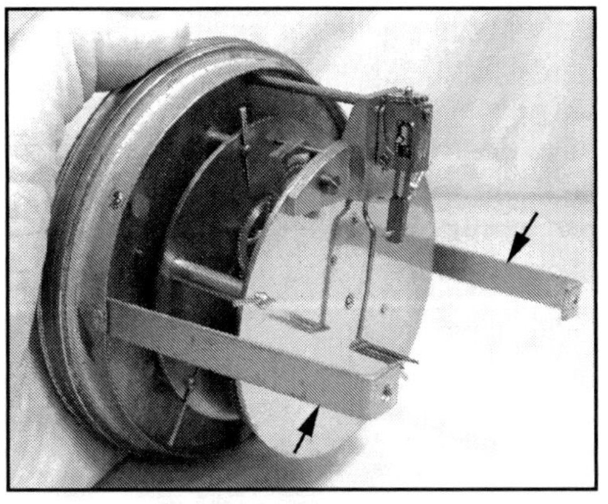

2

RACK & SNAIL VISIBLE ESCAPEMENT CLOCK

This modern replacement case is made of contrasting, natural woods. It has much of the feel of an old case and would grace any mantel.

French movements have a way of turning up surplus, after the heavy marble cases are broken or chipped so badly that they are scrapped. Almost any round French movement and dial unit, with its adjustable strap mountings, can eventually be matched to a different case. Be especially careful about buying or repairing any French clock whose originality is suspect. The movement may have been fitted with a short pendulum that fits the new case but will not keep time. Even if the serial number on the pendulum bob matches that on the back plate of the movement, there can be a problem, since a pendulum rod can be shortened. Instructions on "reading" the pendulum length are found on page 17.

To create our featured clock, someone started with a surplus visible escapement movement with a numbers-matching pendulum. He or she decided to install the mechanism in a custom made wooden case, rather than trying to find an old case with acceptable dimensions.

The so-called visible or Brocot escapement is a pleasing design that adds interest and value to a French clock. Many of the visible escapements have red jewel pallets that add a touch of color to the dial. It is fascinating to watch the pallets swing from side to side.

The dial has a recessed center with a visible escapement. The pallets are pieces of jewel. Unfortunately, the name of the maker or retailer (between the winding holes) is mostly worn away. A part of the word "Paris" is still readable just below the name.

French-style mounting straps were adapted to fit the antique movement to the new case. As mentioned, the pendulum is original to the movement (serial number 3099). However, the gong is not an old one, but a modern replacement.

This French china-cased clock retains its original gong. The gong base is a rectangle of brass with two mounting screws to hold the fitting for the gong. Look for this style of gong in a French clock.

Disassembly

Returning to our featured clock: Unhook the pendulum. Then remove the strap screws at the back of the case and pull the entire movement/dial unit out the front of the case. Next...

Three bezel screws will release the bezel and strap unit from the movement.

Three taper pins will free the chapter ring, which can then be removed.

Let down the power of the two mainsprings next. As shown, the smooth handled let-down key is managed with one hand. A finger of the other hand can depress the click to release it. Allow the handle of the let-down key to rotate slowly in your hand, until the mainspring is fully released.

It may be a safer idea to depress the click with the tip of a small screwdriver. I used my finger here, only to find out a few minutes later that the clickscrew was very loose. The click could have let go! There was no damage to the click or the screw threads; the screw was just loose.

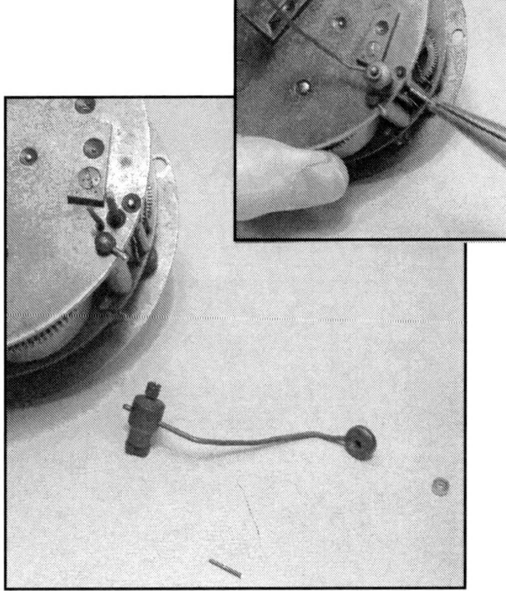

The strike hammer can be removed after the taper pin is pulled out. It is often necessary to pry gently at the hub of the hammer assembly to loosen it.

Remove the screw and then pull off the back cock. This frees up the back end of the pallet arbor.

Turning to the front of the movement, remove one screw holding the pallet cock. Now the pallet unit can come out.

Remove the two screws holding the escape bridge, then follow with the bridge.

The escape wheel and arbor can come out next, but be careful of the fine pivots.

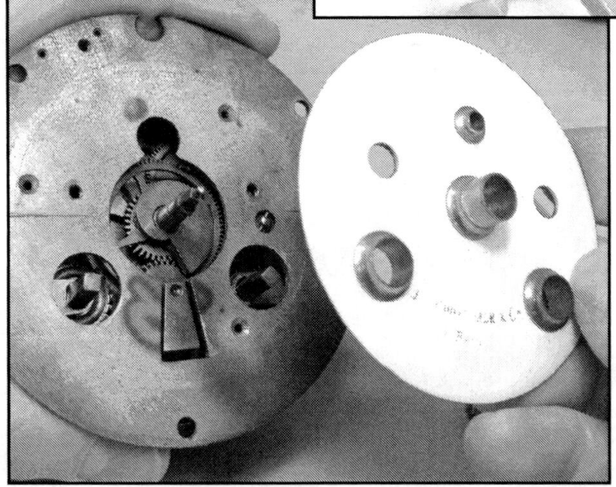

When the white dial center is taken off, a brass false plate is left. This, in turn, is removed after another set of taper pins is pulled out.

The back plate carries several markings that provide information for the repairer. It is a good idea to look at these clues. You will then have the best possible chance of repairing the movement properly. You may be able to spot an incorrect pendulum or a bad "second marriage" clock—one with a non-original movement that does not fit the case—before you start the repair. If the movement requires a pendulum that is too long for the case, your repair is not going to work out.

Look for a serial number. Our featured movement is marked "3099" on the back plate. The same four digit number is found on the pendulum bob, showing that it is original. The dial false plate and the white dial center are each marked with "99", not the full number.

The pendulum length, stated in "pouce and ligne" measurements, is stamped at the bottom of the plate (see explanation below).

The trademark is a stamp from Marti, and it is a "Medaille de Bronze" or second prize. In his book *The French Marble Clock,* author Nicolas Thorpe links this mark to the Paris Exhibition of 1855. This shows that the movement was manufactured after that date.

Two removable cocks, one for the center arbor rear pivot and the other for the strike pinwheel arbor, are indicated in the photo. These are further marked with dot punch marks, one for the time and two for the strike. Even the screw heads are marked.

Pouce and Ligne

The numerals stamped at the lower edge of the back plate, on either side of the pillar, represent the pendulum length stated as follows:

- Numeral(s) to the left of the pillar = "pouce" (one pouce = 27mm)
- Numeral(s) to the right of the pillar = "ligne" (1 ligne = 1/12 pouce or 2.256mm).

In this movement, 4 pouce = 108mm and 8 ligne = 18mm. The length is 108 + 18 = 126mm or 4.96 inches. This is the distance from the bending point of the suspension spring to the center of mass of the pendulum bob, located just above its center. The actual length to the *bottom* of the pendulum bob is greater and is necessarily an estimate. In this case the original pendulum measures 141mm or just over 5-1/2 inches.

Remove the minute wheel cock.

The motion work is taken off: the
minute wheel, hour wheel &
snail unit, and cannon pinion.

The nicely curved strike
rack is removed next.

This movement has an
external time wheel that
drives the escape pinion.
Remove the bridge and
then the wheel.

Remove the taper
pin, the washer, and
then the rack hook.

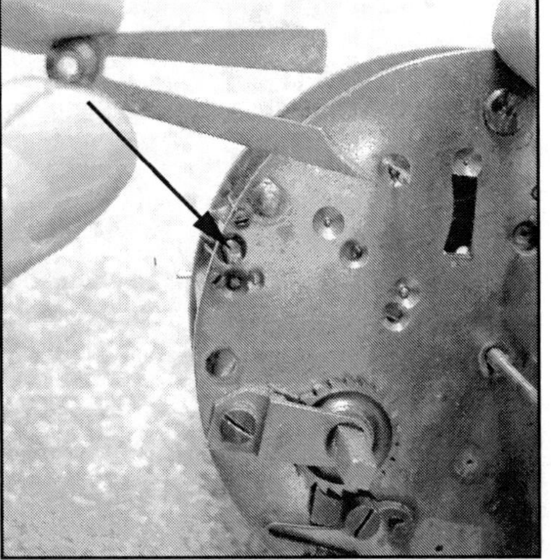

The lifting/warning
lever assembly is
next.

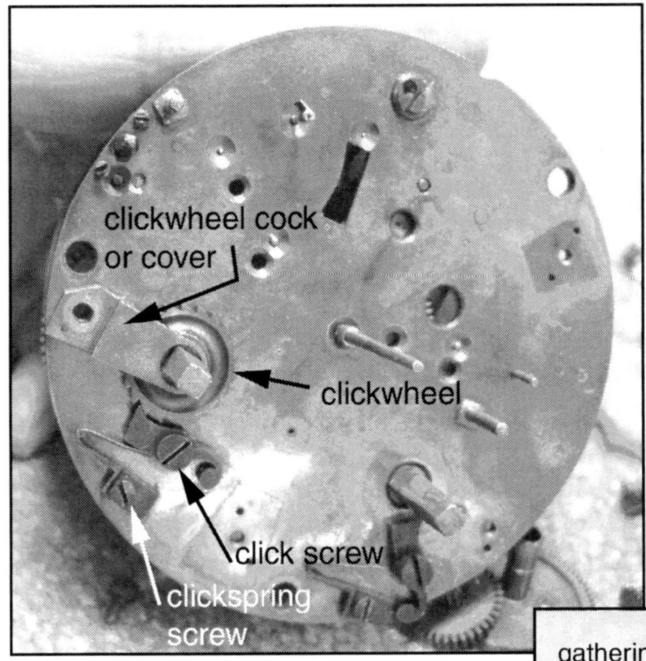

The click assemblies should be completely removed before the movement is cleaned. That means the clickspring screw, clickspring, click, and clickwheel should all come off. Leaving the parts on the movement will trap cleaning solution or oil under them, making a mess. The clickscrew indicated in the photo is the one that I found very loose.

The strike side winding parts are featured in this photo. Note that the clickwheel cover retains the clickwheel. The cover on the strike side normally has a beveled edge, as this one does. This may allow a dial foot to clear in some movements; otherwise it is just a feature that makes it easier to identify the strike-side cover.

Remove the gathering pallet from the front of the movement. The pallet is tapered; it is a "push fit" on the long pivot. You can try a gentle twisting, pulling motion to remove it, but if it does not come away, do not risk breaking the pivot. Leave it until the plates have been separated. Then the front plate can be supported and the end of the pivot gently tapped to remove the pallet. Put this small part away in a secure place. You will not want to make one.

Pull out the taper pins that hold the plates together.

The movement looks like this with the back plate lifted away. Remove arbors carefully to avoid bending the smaller pivots.

On this movement, the visible escapement utilizes a cock to support the rear pivot of the time fourth wheel. Take off the screw and then the cock, as shown in these views. Remove the center arbor.

The Pivots

From looking at the outside of the movement, I had already identified the center arbor front pivot hole as a candidate for a bushing. But now, as I looked at the center arbor rear pivot, I saw that it was worse than that; the pivot was severely worn. The arbor was set up in the watchmaker's lathe, supported by a steady rest. The scored pivot was turned to a slightly smaller diameter and polished.

After both center arbor pivot holes had been bushed, the arbor was installed between the plates to test for free running. The fit of the pivots in the holes was made to a finer standard than an American clock; special care is needed to achieve the correct fit. Spin the assembly and make sure it will continue spinning when the front plate is oriented up, then down. Shake the movement and verify that the arbor clicks forward and back to the limits of the endshake. There is another test: look at the pivot in its hole and move the arbor. With a loupe, you must be able to see some slight side to side movement of the pivot in the hole, or it is probably too tight.

A graver will cut the hardened steel of the pivot.

A pivot can be stoned, as shown here, instead of cut with a graver, but a graver is still needed to put the sharp corner back on the shoulder, where the pivot joins the arbor.

The rear pivot of the time second arbor was found to be badly scored. Just like the center arbor rear pivot, it was dirty and had probably been run for years without attention. (This is a typical wear location on a French movement, a high stress point with the wheel and pinion located right at the end of the arbor.) The unworn portion at the end of the pivot was .058" diameter, but at its deepest groove, the pivot measured about .053". At the same time, the pivot hole had not worn noticeably oval.

The hardened pivot must be examined to see if it can stand the reduction in diameter that is required to restore the pivot to a smooth cylinder. I thought this one looked promising. (If it had been too badly worn, the best course, without carbide drills or other special setups, would be to send the wheel and arbor to a specialist for repivoting.) After stoning and some graver work, followed by polishing, the finished pivot measured .052" in diameter, a reduction of .006".

As a next step, the pivot hole needed to be bushed; it certainly had grit embedded in the wall. This would have quickly cut into the newly polished pivot.

The hole is bushed in the Keystone Bushing Tool.

A ball shaped cutter in the bushing tool enlarges the oil sink in the new bushing.

A "spin and shake" test determines that the new bushing allows the pivot to turn freely.

Removing the Barrel Cover

Select a screwdriver that will fit in the slot in the cover. The blade must be strong enough to resist bending. I have a favorite small screwdriver (shown) that works well for many small screws and for the task of loosening barrel covers. Some will argue that the edge of the barrel may be bruised by the prying operation. I have not found this to be much of a problem, especially as most covers have been removed this way in the past and are already slightly marked. This is my preferred method.

Another way to remove a cover is to tap the back end of the winding arbor against a block of wood. The front shoulder of the winding arbor pops the cover loose. Sometimes the cover does not come loose with a reasonable tap; tapping harder will bow out the cover, requiring repair.

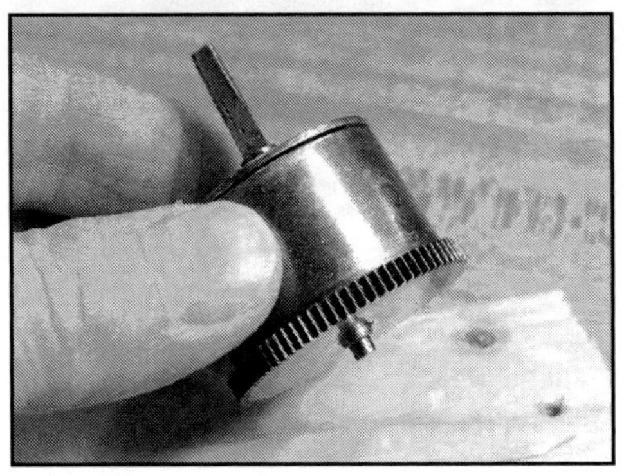

Extracting a Mainspring From a Barrel by Hand

Removing a French spring with a winder (page 53) is safer and easier than pulling it out by hand. However, if you do not own a winder, hand removal and insertion is possible with most French springs, as they are relatively weak.

Wear heavy gloves (not shown in the photo below, to enhance the clarity of the image). Wear safety glasses! Pull out the center coils with pliers as shown. Grasp these coils with your fingers and begin pulling out one coil at a time. Finally, unhook the hole end of the spring from the barrel hook. There is a risk of some distortion of the spring into a helix shape.

From an examination of the spring that was taken out of the barrel, I found that the outer end appeared to be original and it was not split or rusty. The mainspring was reused after cleaning.

It is common to find mainsprings with cracked or broken-out hole ends. It is sometimes possible to use a round file to remove a 1 millimeter long crack that is just starting. But if the damage is more severe, you will have to either shorten the spring or replace it. If a spring has already been shortened, it cannot be shortened a second time. See page 8 for details on how to shorten a mainspring.

Next to the spring, the photo below shows the other spring from our clock still in its barrel. This view gives an indication of how "full" of mainspring a French barrel should be.

Installing a Mainspring in a Barrel by Hand

Hook the outer end of the mainspring onto the barrel hook or rivet first...

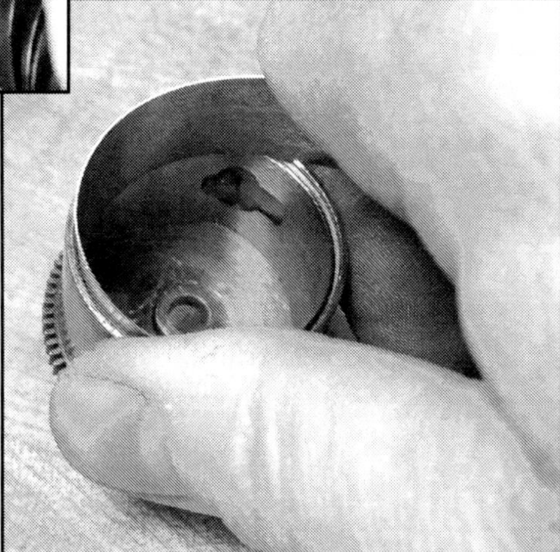

...and, as you wrap in the first coil, make sure the hole end is still firmly hooked onto the barrel hook.

Carefully wind in one coil at a time, retaining each one as you go.

Keep going until most of the coils are in the barrel. I always need to pause at this point to relieve cramps in my tired hands!

The inner coils can be twisted a little and popped into the barrel. The task is almost complete.

The last step is to insert the winding arbor into the barrel. The arbor hook must be safely hooked in the mainspring inner eye. Before you insert the arbor, form the inner eye with pliers to obtain a good fit.

With the mainspring in the barrel, add a small amount of mainspring oil. Some repairers will lubricate the springs before inserting them, but I find it hard to handle the slippery springs safely.

The notch in the barrel cover usually lines up with the barrel rivet on the outer wall. Many of the barrels also have a mark for alignment. French barrels are so well made that many of the covers can be snapped in place with the fingers.

If a little persuasion is needed to seat the barrel cover, you can use a small, smooth-jawed vise and squeeze gently. If the cover does not go into place easily, do not force it; step back and try to discover what is wrong. For example, is this the right cover for the barrel?

Polishing

A book could be devoted to polishing. Most French plates are not lacquered, so a polished clock plate will darken with the years. By the time it really needs polishing again, the movement needs cleaning, anyway.

Metal polish may be slow to work. I use diamantine powder mixed on the plate with oil, rubbed and buffed with a cloth. This cuts through the corrosion very well. Twirl a stick covered with a scrap of cloth in each oil sink. Carefully remove all polish residue by recleaning the plates. Pay particular attention to the pivot holes.

One movement may not receive any polishing, another might have just the back plate done, and a third might have all parts, front and back, laboriously hand polished and all screws reblued. It will depend on the cost of the repair or the care with which a hobby clock is restored.

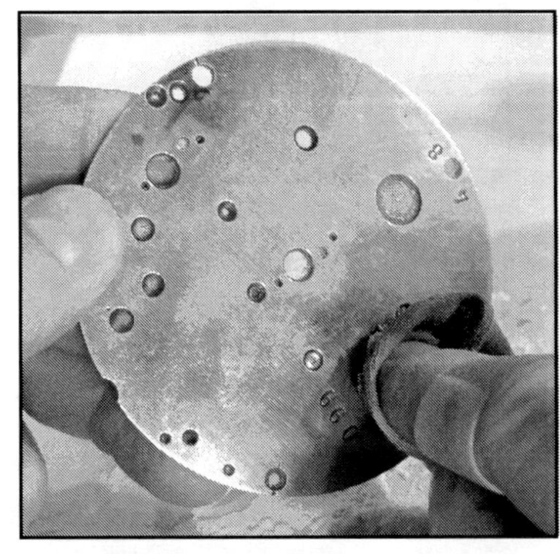

Reassembly

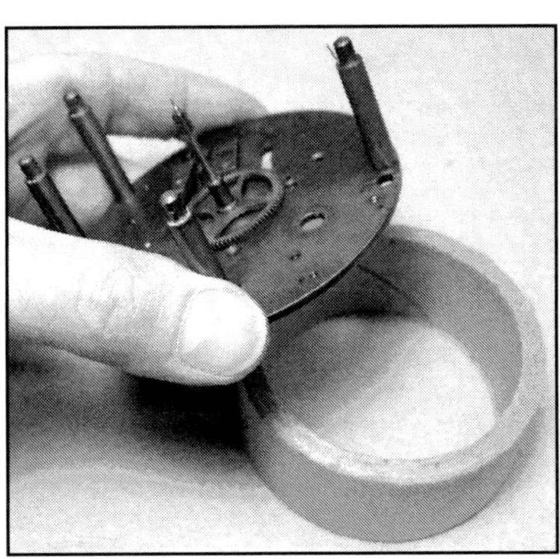

Following the cleaning of all parts, reassembly begins with the front plate. Find a support that will allow the plate to lie flat, but raised above the bench so the center arbor will clear. I have several cardboard cores that were cut from a longer tube. These cores make good movement assembly stands. There are also special clamps made for the purpose. Even a small cardboard box may fill the requirement.

This movement has two cocks mounted to the inside of the front plate. Install the center arbor first and then add the cocks (indicated by the arrows). The one on the left supports the rear pivot of the time fourth wheel. The one on the right supports the escape arbor rear pivot.

Add the mainspring barrels next. The barrels are usually marked by the maker. One dot identifies the strike in our movement; there is no dot marking on the time barrel. In other French movements, two dots stand for the strike and one for the time. Barrels may also have been scratched "T" and "S" by a repairer, but such identification cannot be relied upon.

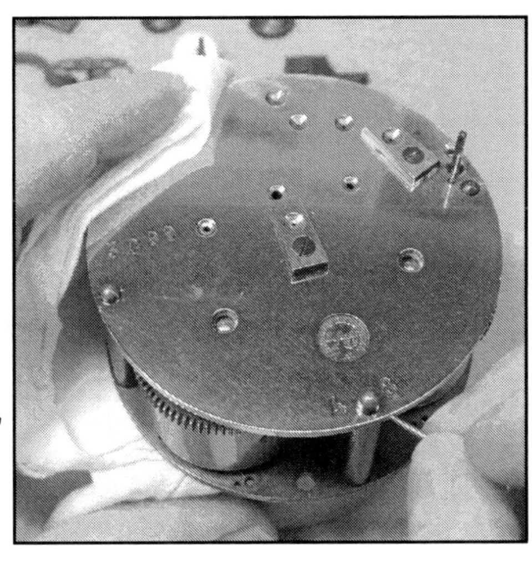

After the back plate has been added, lightly install a taper pin in the bottom pillar. This will help to prevent the plate from "see-sawing" after you have placed some of the pivots and try to move others into their holes.

Once the plates are together, install taper pins on all four pillars. Then add some of the front movement parts that are needed during the adjustments to the strike mechanism.

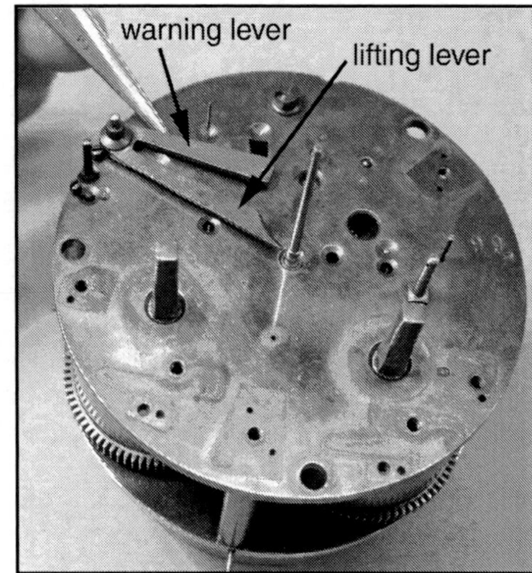

Add the lifting and warning lever assembly. The assembly pivots freely on the post.

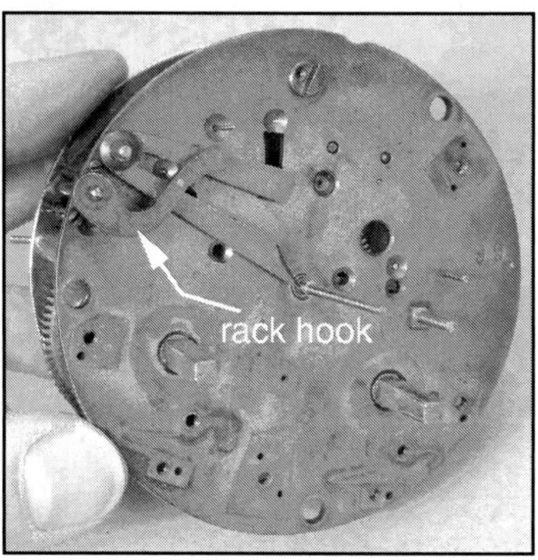

Now add the rack hook. It is placed over the lifting/ warning assembly and fits over a squared arbor. There is a steel pin on the rack hook which must be located between the lifting and warning levers.

Fasten the lifting/warning lever assembly and the rack hook with taper pins. Start with new steel pins. Insert them, mark the lengths, remove them, cut the ends, and file the ends smooth. Rough-cut ends or bent pins are not the mark of a good repair. The rack hook has a brass washer under the pin.

Setting the Hammer Tail

The hammer tail should be adjusted so that the hammer is at rest when the strike train is locked. This is accomplished by adjusting the mesh of the pinwheel and the pinion on the gathering arbor. It can be done without separating the plates.

Here's how it is done. Mainspring power must be completely "off". Remove the pinwheel cock from the back of the movement; this allows the pinwheel to be moved out of mesh and readjusted. Rotate the wheels until the lock pin rests against the locking lever, as shown in the photo at the right. At this point, you want the hammer tail to be free of the lifting pins on the pinwheel. Many French clocks, including this one, are marked for this position. There is an obvious punch mark on the wheel, placed between two teeth. There is also a leaf of the gathering arbor's pinion that has been ground at one end. I couldn't seem to see this pinion mark— that is, until the movement was assembled as shown. Then the pinion mark was clear! The light caught the corner of the pinion leaf, making it show plainly. I eased the pinwheel clear of the pinion, rotated the wheel, then meshed the marked tooth space on the wheel with the marked pinion leaf. The hammer tail was free of the pinwheel pins: this was the correct adjustment. I reinstalled the cock on the back plate.

Setting the Warning

The next adjustment is made to the warning wheel. Rotate the gears until the lock pin again rests against the locking lever. We know the hammer position is correct, since we just set it up in the previous step, but what about the position of the warning pin? The pin should rest about one half revolution away from the warning lever. As the clock "goes to warning" before each hour and half hour, the pin and wheel rotate this half revolution. This translates to an 11 o'clock warning pin position, viewed from the front (for most French rack strike movements). The warning "run" can be less, that is, close to a quarter revolution, but it should not be more than a half revolution. Too long a run may allow the pinwheel to begin raising the hammer tail. In fact, the hammer tail should remain free before and after the warning. When the strike begins, the gears should reach full speed before the hammer tail is lifted. This helps to prevent stalling.

First...

Check that the mainsprings are still let down! Remove two of the four pillar pins and separate the plates by hand or with a plate separator, as shown. This is a partial separation. Do not take the movement completely apart! The plate separator allows you to move the plates apart exactly as far as you desire, until a particular pivot can be eased out of its hole and the gears meshed differently. The device has a spring which gently keeps the plates from moving further apart than is desired.

Then...

Grasp the warning wheel with tweezers and move it to the front, so that the rear pivot comes out of its hole. Now you can re-mesh the warning wheel with the pinion on the gathering arbor.

Repairing the Brocot Visible Escapement

What caught my attention, even before I repaired this movement, was the narrow arc of the pendulum swing. Laurie Penman, in his book *Practical Clock Escapements*, associates narrow arcs with relatively longer center distances between the pallet arbor and escape arbor. Our movement, with its 36 fine teeth, has a long center distance, or height. The narrow swing explains why the clock stops when it is slightly out of beat.

The pallets on these clocks are tiny cylinders of steel or jewel ground down to the diameter, to create D-shaped pieces. I found, initially, that the flat plane of the entry pallet (on the left side) pointed far down, rather than more towards the center of the escape pivot, as is correct. The pallets were held in place with shellac; it was necessary to heat the shellac to soften it enough to allow the pallet to be twisted in its socket.

An alcohol flame can be used to heat the shellac that binds the pallet jewel in its socket. Once soft-ened, the shellac stays warm from the heat retained in the brass pallet body.

Here is another heating method which works well. A clean soldering tip with no solder on it can be used to heat the pallet body.

Whatever the heating method used, there is enough time to gently grasp and rotate the pallet to another position. It can be a challenge to keep the pallet upright, if the hole in the body is excessively large.

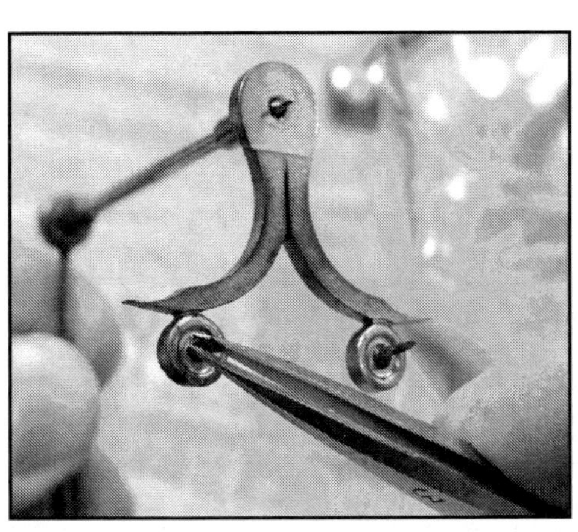

The diameter of the pallet is typically 90% of the escape wheel tooth pitch, but it can vary. (This is stated in Laurie Penman's book *Practical Clock Escapements*.) I measured the pitch (the distance between teeth) in our escape wheel as .070". At 90%, the pallet diameter would be .063". I also measured the existing pallet jewels as .066"; they were ground down to .031" at the flats. I checked a supply catalog and found that the jewels from the movement could be replaced for almost $10 each. Further, the sizes .066" (the current jewel size) and .063" (my calculated size) were both listed.

To make new pallets, I located a 1/16" diameter (.0625") rod of unhardened tool steel. The first practical problem was that the existing jewel holes in the pallet body were .071". I have never liked loose fitting jewels very much, since it is awkward to keep them standing upright in the slowly hardening shellac. Since I was changing to steel pallets, I decided to install bushings in the pallet body holes. I bushed the holes to .059" and then carefully broached them out to an easy fit for the steel pallet stock. This would allow me to twist the pallets in their holes later, and the pallets would always remain upright after an adjustment.

Making the steel pallets begins with the polishing of the steel stock in the lathe. Then file the "flat", measuring as you progress to half the diameter. Finish by stoning the flat surface. Remove any burrs at the edges. After cutting off each pallet from the parent rod, I heated it to red in a propane flame and then dropped it into water, leaving the steel dead hard.

Stoning the flat on the new steel pallet

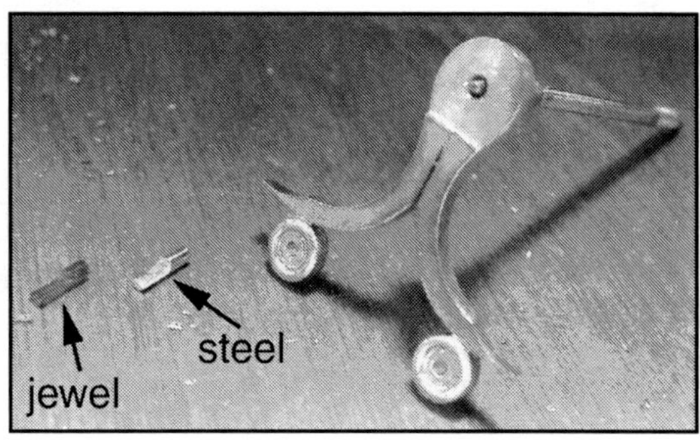

This photo compares a
steel pallet to a jewel pallet.

Here is the finished pallet assembly.

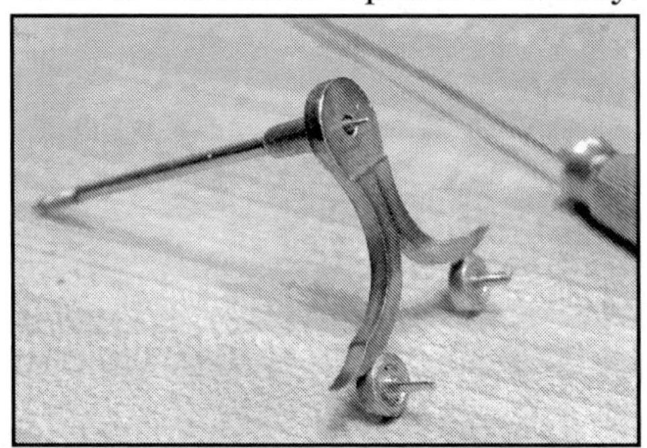

Why did I replace the pallet jewels in this movement? A few moments after I took one of the photos, a cardboard backdrop I was using on the bench fell over, struck the pallet arbor, and knocked it out of the vise which held it lightly. One pallet jewel broke off! I had just earned myself the task of making or replacing the pallets. I decided that making steel replacements would be more helpful for our purposes. However, new jewels are available, and they can be installed fairly quickly using the techniques illustrated.

There are two potential escapement changes or adjustments to consider together. First, the height of the pallet arbor may have been changed. Most French escapements have an eccentric adjustment, but just because it is there, do not assume it will be easy to use. The eccentric has a screwdriver slot that may have been damaged, making it impossible to move. The eccentric can be tapped out through the front of the pallet cock with a flat punch and hammer. Then the eccentric can be put gently in a vise and the slot treated with a screw head file. Now ready to go back in place, the eccentric is installed and a small, lower chamfer may need to be tapped back in place, if it has been disturbed. In our clock, the eccentric is smaller in diameter than the hole in the cock. That is the reason the lower chamfer is there; without it, the eccentric would simply fall out. After this work, the eccentric now turned easily with a correctly sized screwdriver.

The second potential change is not an adjustment, as such. Someone might already have squeezed the pallet body closed in a vise or opened it up by prying. This changes the distance between the pallets (the span). Generally, if the pallet body bears no marks and appears to be original in shape, do not alter it.

To verify the height and span of the pallets, check the following, as shown in the drawing. Move the crutch to one side, so that a pallet (the exit in this drawing) has an escape tooth locked on its diameter. At the same time, the other pallet should be just touching the outer diameter of the escape wheel. This is the diameter at the tooth tips. The span should be exactly right, and if it is not, then some change has been made to the height or span. I ended up by turning the eccentric to slightly reduce the height. Span and height are related, so a change to one affects the other.

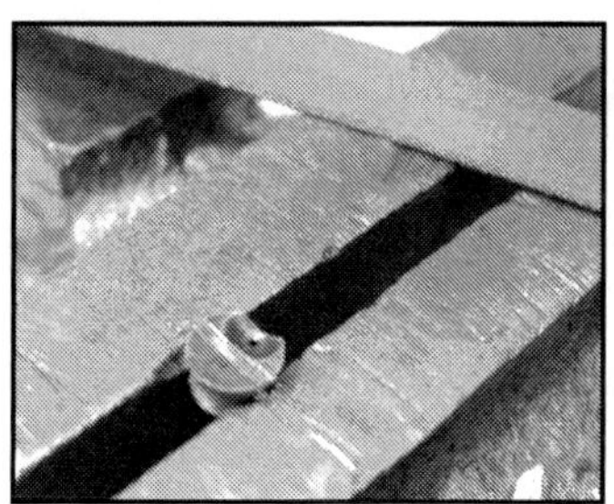

The eccentric is held in a vise so that the screw slot can be "cleaned up".

The eccentric is ready to be installed back into the pallet cock.

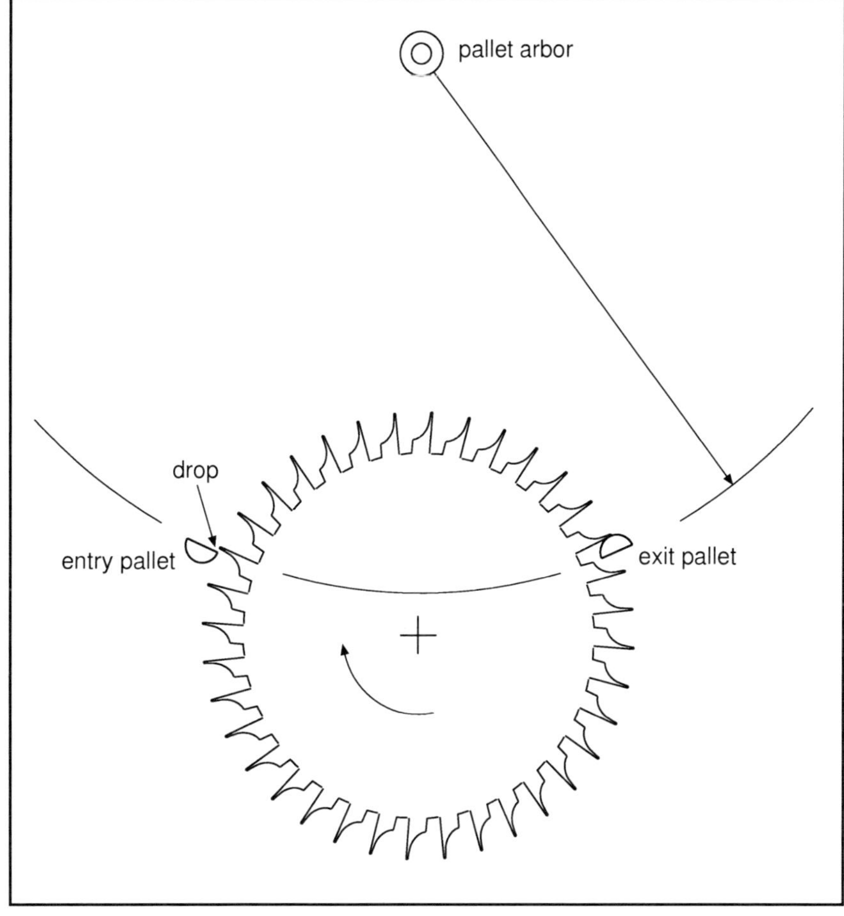

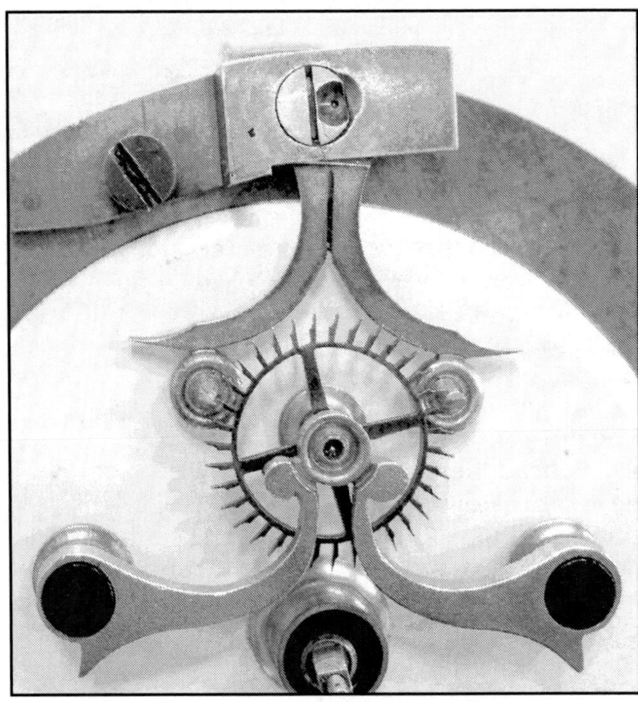

The finished escapement is shown
before the chapter ring was reinstalled.

Adjustment of the Brocot escapement can be difficult. From my own experience, it seems that things should be taken in order. Escapement pivots and pivot holes should first be determined to be in good order. Then the pallets should be verified for size and condition. Twist the pallets in their sockets (page 29) so that the "flats" point to the escape arbor center as a starting point. Whether the pallets are steel or jewel, chips, worn edges, or grooves are a cause for replacement with new pallets.

The next step is to deal with the related dimensions of the span and height, as described on the previous page. A final step is to discover whether the pallets need to be twisted again in their sockets, as a fine-tuning to adjust and equalize the drops. I have usually found that the twisting of the pallets makes a difference in whether a pallet will catch the back of an escape tooth or not, rather than producing a fine adjustment of the drops.

After all the work was completed on this French escapement, I found that its condition was improved. The flats of the pallets now point more towards the escape pivot center than before. In addition, the escape teeth, which failed to lock on the diameter of the exit pallet, now do so. I can conclude that the span may have been altered and the height needed adjustment. I am also convinced that the slightly smaller diameter pallets I made have smoothed the operation of the escapement.

After cleaning and repairing the movement, including the escapement, I find that the pendulum swing is still small, as determined by the design of the 36-tooth escapement. The clock must be set accurately in beat if it is to run.

As I explained at the beginning of this chapter, the movement was housed in a custom made wooden case. The visible escapement made it a standout as a project. However, it is easy to believe that the clock may have been set aside when the movement failed to perform to expectations. The escapement is a fine one that requires careful repair and adjustment.

3

COUNT WHEEL STRIKING CLOCK

One of the most common French clocks is the round movement style with a simple rccoil escapement, a bell on the back of the movement, and a count wheel strike mechanism. Case styles vary, with black marble and ornate cast metal cases predominating.

Our example has an ornate cast metal case. Fortunately, no one had ever given in to the temptation to coat it with gold paint. The original plating cleaned up rather well, revealing an original, unmolested appearance under the dust. I used a damp brush to clean the surface, avoiding any polish. The dial is in good condition, with only the most minor chipping around the dial holes.

The clock was brought in because it did not strike. A quick application of the winding key indicated a broken or unhooked mainspring. The movement was also very dirty, indicating that it had not been serviced for many years. However, once I looked closely at the movement, I found evidence of skilled bushing work, always a good sign.

The clock's attractive enamelled dial states the dealer's name and address.

This is a rear quarter view of the Marti movement.

The ornate case is 19 inches tall and 12 inches wide.

The Broken Strike Mainspring

Close inspection of the strike wheels and pinions did not reveal that any damage occurred when the mainspring snapped. The break in the spring can be seen in the barrel. The pieces were pulled out in the manner described on page 22. Extra care is always needed in the removal process because of the jagged ends.

Choosing a Replacement Mainspring

1. Measure the broken spring. This one is 20mm wide x .012" thick x 48" long.
2. Check catalogs to see what is available:
   - 19mm x .011" x 49" (use as-is) or
   - 19mm x .012" x 53" (shorten to fit)
3. Try your selection; I chose the .011" thick spring, on the right in the photo. I installed the spring in the barrel, placed the barrel on the winder, and found it could be wound 6 turns. This compares favorably to the 5-1/4 turns produced on the time mainspring barrel. I felt this choice would work.
4. Test the spring in the clock, to make sure the reduction in thickness from .012" to .011" will not affect the striking speed or the number of days the clock will strike. This test was successful for a one week run.

Two possible replacement springs
are shown side by side.

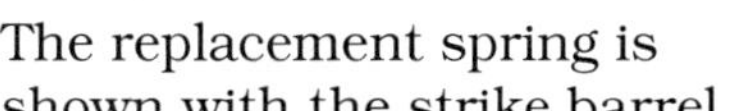

The replacement spring is
shown with the strike barrel.

Tightening the Barrel Hook

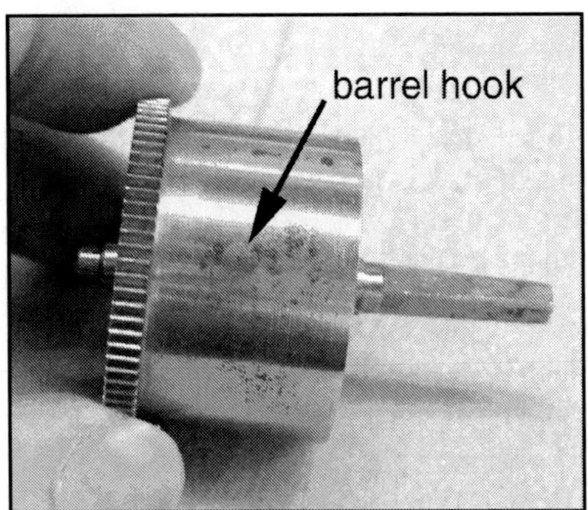

The barrel hook is a specialized rivet. Generally a tight, correct hook will be flush with the outer wall of the barrel, as shown in the photo. The steel rivet blends right in with the smooth brass of the barrel.

The part of the rivet that extends into the barrel will have a hooked end or a cap that extends all around it. This will securely hold the end of the mainspring.

This cutaway drawing compares a loose, bulged-out barrel hook to one that is tight and flush with the barrel wall. It is difficult to hook a mainspring onto a loose hook, and it is likely to allow the mainspring to slip off.

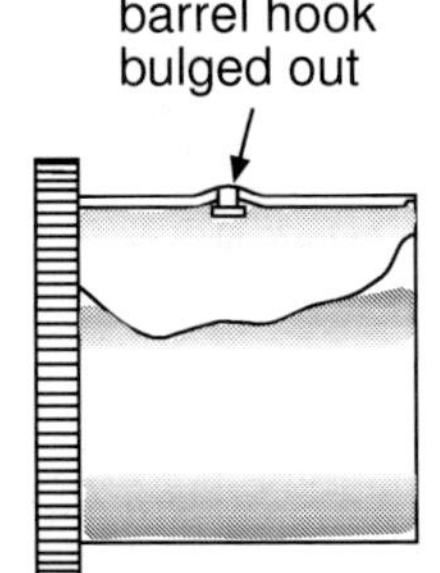

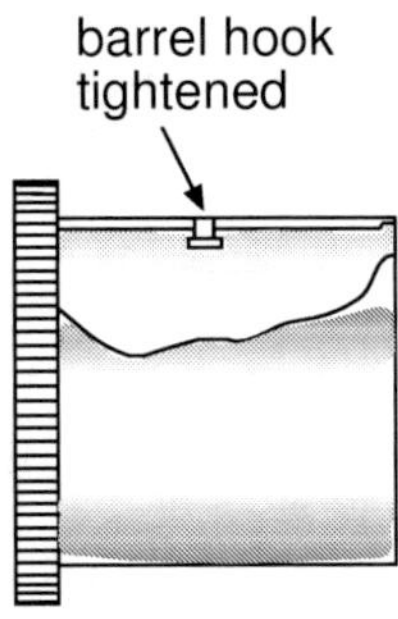

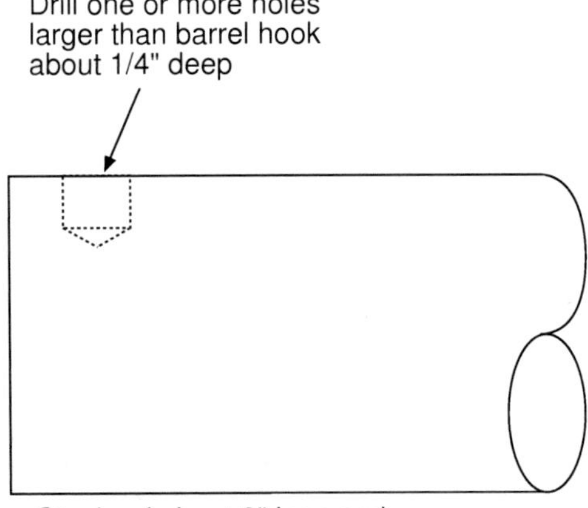

To tighten a loose barrel hook, make an anvil similar to the one shown in the sketch. Clamp the anvil firmly in a vise. Place the barrel over the anvil and line up the barrel hook with the hole in the anvil. Tap down on the hook with a hammer to push it inward and tighten it. Take extra care to avoid striking the teeth of the barrel wheel.

Assembling the Parts

The parts are shown arranged on the bench top with the time parts on the left and the strike parts on the right.

Begin by placing the front plate face down on a support such as a cardboard box (or use assembly clamps). This will permit the center arbor to be installed with the long front portion extending downward. This is followed by the two spring barrels.

Add the strike hammer arbor, followed by the strike locking and count lever assembly (shown in the background, on the bench top).

Install the rest of the parts. It may be helpful to leave out the delicate fly, as shown, until the back plate has been added and some of the rear pivots are in their holes.

Install the back plate lightly. Do not force any pivots into place. The first pivots that will go "home" are the extra long ones: the hammer arbor and the long, squared arbor for the count wheel. The next ones to seat are the winding arbor pivots.

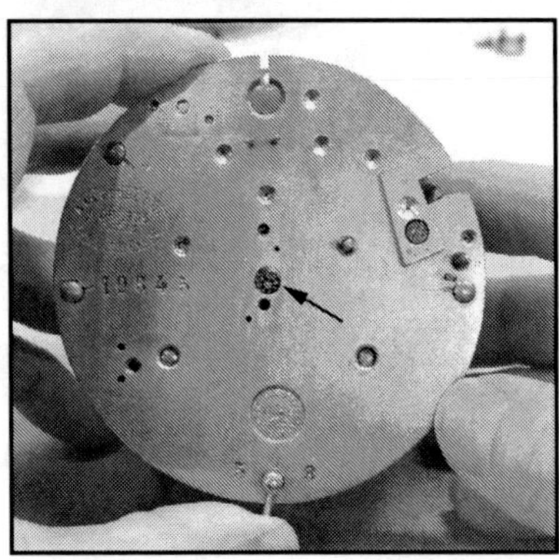

The center arbor rear pivot should also be one of the first to find its pivot hole. However, in this clock I couldn't seem to guide the pivot into the hole, and I was careful not to force it. Perhaps there was an irregularity on the surface of the center arbor cock. The answer was to remove the cock (see arrow), just until several more pivots were in place. Then I installed it again, placing the center arbor rear pivot into the hole without a problem. Install the lower pillar pin now, by hand. This will prevent the rear plate from tilting as you fit the pivots.

Use tweezers to gently move the arbors around and find each hole. If you are careful, there is little chance that any will be bent during the assembly. Insert the other pillar pins as it becomes possible to do so. Use the old pins for now or replace them with new steel pins trimmed to length. It will have to be done before the repair is complete.

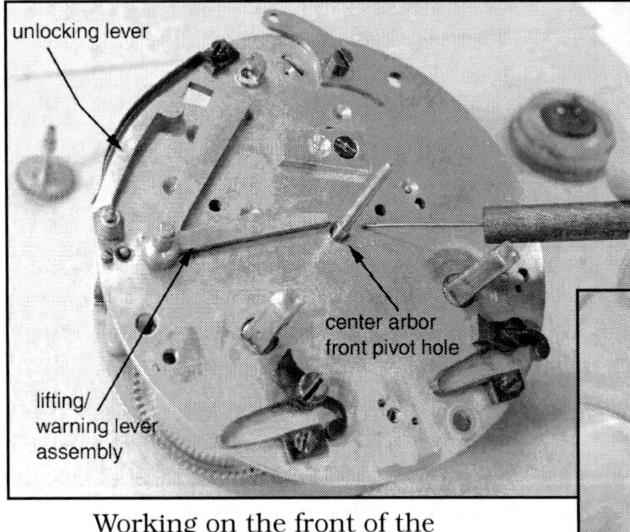

Working on the front of the movement, begin to install some of the parts as shown.

This view shows a good opportunity to grease the lifting pins on the back of the cannon pinion, before it is installed on the center arbor.

Synchronizing the Hammer Tail Position

Synchronizing the strike hammer tail position is required so that the hammer is not left "on the rise" when the strike sequence is completed. Begin by identifying the pinion on the locking arbor (below, left) and the punch-marked tooth space on the pinwheel, which meshes with it (below, right).

After the movement has been assembled, the same parts and their marked places are easy to find (above).

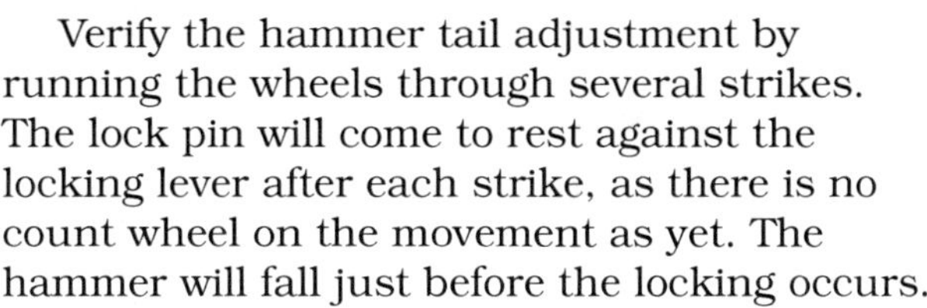

With the mainsprings still completely let down, remove the pinwheel cock. Now you can slide the pinwheel arbor to the rear, disengaging the pinwheel from the locking arbor pinion. Set the marked pinion leaf in mesh with the punch-marked tooth space, and the hammer tail is synchronized.

Verify the hammer tail adjustment by running the wheels through several strikes. The lock pin will come to rest against the locking lever after each strike, as there is no count wheel on the movement as yet. The hammer will fall just before the locking occurs.

Setting the Warning

The warning is the second and last between-the-plates adjustment to make. The photo at the right shows the lock pin against the locking lever. The hammer tail is clear, because we have already set the marked wheel and pinion (page 39).

At this point, the warning pin should rest with a quarter turn, but not more than a half turn, of wheel rotation ready to occur when the mechanism is unlocked for the warning.

If one views the movement *from the front,* the warning pin should rest at a 6-8 o'clock orientation. This is not the same as the setting for the rack striking clock featured in Chapter 2.

The warning must be set by meshing the warning arbor pinion to the locking wheel. The plates must be separated just enough to allow the adjustment.

Needless to say, there is still no power on the mainsprings! Pull out three of the pillar pins and ease the plates apart slowly. Use a plate separator (below) for the best control.

This photo shows the path of the warning pin as the warning occurs.

Care must be exercised to avoid stressing any of the fine pivots. Check the meshing of the gears by verifying the position of the warning pin, when the lock pin rests against the locking lever.

It is possible that the warning run will fall into an acceptable range without having to adjust it; it's a matter of chance in the assembly of the movement. Adjusting the warning takes only a few minutes, and it is an important adjustment.

Nonetheless, it is all too common to find movements that are assembled at random, with the hammer stopping on the rise or a warning that is too long or too short.

What is the Warning?

Warning is the motion that makes a clock ready to strike. Most, but not all, strike mechanisms have a warning. The Morbier wall strike clock covered in Chapter 5 is an example of one that does not.

The warning wheel has a pin in it which moves through a set arc when the strike is first unlocked, about five minutes before the hour and half hour. This pin holds back the strike gearing until the exact time for the strike to occur.

Why have a warning at all? After each strike sequence, the train locks heavily and positively. Beginning several minutes before the next striking point, the movement slowly lifts its way out of this "lock-down". It is then that the warning system holds back the train—much more lightly—until the moment for the strike. Then it takes little force to unlock the strike to run.

Finishing the Assembly

Add the clickwheels, their springs, and the clicks. It is common to find damaged or replaced parts here. On this movement, I found that one click was a very loose fit on the clickscrew. I looked through some spare parts at a dealer's shop, and I came to a conclusion. It is much easier to find another click to fit your clickscrew than it is to try it the other way around. The nonstandard screw threads in the French movements are the reason this seems to be true. The "new" click fit the existing screw very well.

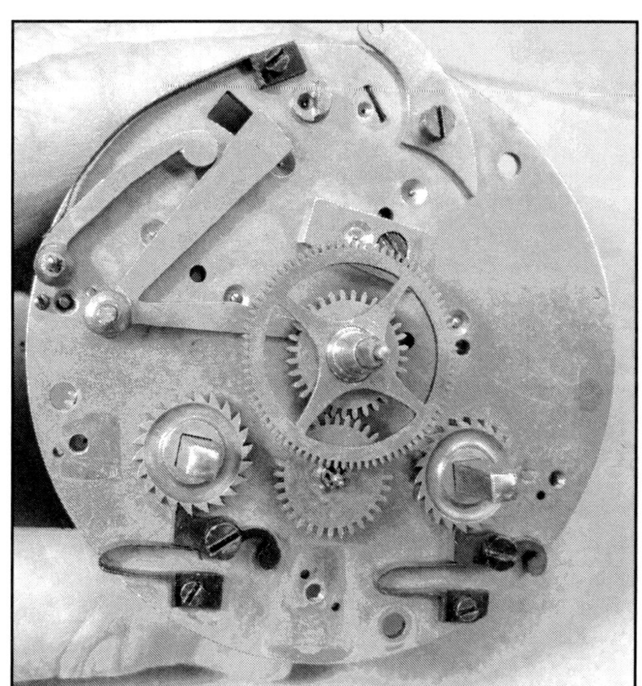

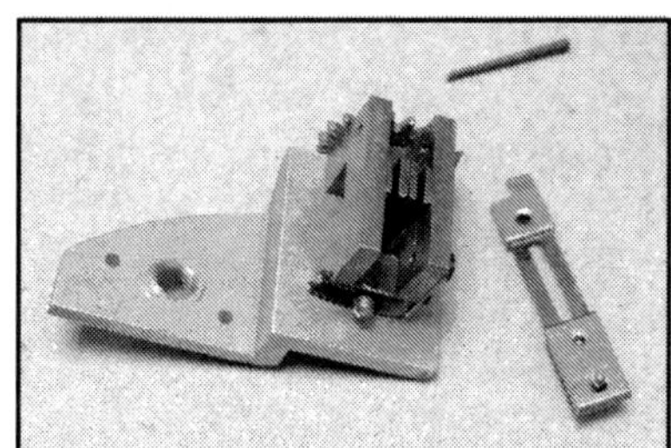

The Brocot style regulator requires the type of French suspension spring shown. The little tab on the top block of the spring prevents it from tipping out of the slot in the jaws of the regulator. Always take the regulator off the back cock for cleaning. Fit a new taper pin with neatly trimmed ends.

After the verge is installed, the assembly is complete. Don't forget the regulator arbor, shown at the very top of the movement. This pierces the dial at "12".

The count wheel may cause some problems. Observe the direction the taper pin fits into the hole in the square shaft in your clock. Install the count wheel consistent with this, but leave the taper pin off for now. The photo shows the count lever resting in a count wheel slot following the half hour strike. There is a small amount of clearance to the left of the lever, as indicated by the arrow. Run the gears through several strikes (ideally through all the hours) and check carefully that there is no miscounting. You can always take off the wheel and rotate it 180° before installing it again. Trim up a new steel taper pin, and above all, do not press it in with extreme force. Many of the holes drilled in the French arbors are off center! Don't risk splitting the arbor.

Check that you have replaced all the pillar pins. Use steel pins. Cutting them to length as shown and smoothing the ends makes for a professional job.

Note the Marti trademark dated 1889.

Oiling

One of the last steps in the repair is to oil the movement. A few parts, such as the center arbor front pivot, should be oiled before other parts are installed over them. It is a good idea to test run the movement for a few hours before oiling the rest of the pivot holes. (If the clock has been oiled and must come apart for an adjustment, the oil will get all over the plates, and they must be cleaned again.) I recommend a good quality, light, natural clock oil rather than a synthetic. Use an oiler to add lubricant to each pivot hole, but do not overfill them. Place a drop on each pallet, on the tips of the clicks, and the places where the clicksprings rub against the clicks. Clickscrews should also be oiled. Lightly grease the hammer tail, strike lift pins, and crutch.

Once the dial is fastened to the movement, install the case screws. This movement does not have a loose, threaded crutch, so the crutch must be bent to set the beat of the clock. It is risky to bend the crutch with it installed in the movement, let alone with the movement still in the case. The reason: the pallet arbor rear pivot may be .025" in diameter or even smaller. Finally, twist the entire movement slightly to fine tune the beat, then tighten the screws.

The finished clock is ready for a test run. I would say that two weeks is the minimum. Some clockmakers run the finished clocks for months; others only test them overnight!

4

COUNT WHEEL PICTURE FRAME CLOCK

The ornate door of the French picture frame clock is 23" tall and 18-1/2" wide.

Our featured wall clock is a common type sometimes called a picture frame clock. Lift up the tall, top-hinged door, and you'll find a six-sided case, just 17" tall, that also supports the movement and gong. Like some American walnut and oak kitchen clocks, this French clock is a simple pine box with a showy front that is meant to impress. With its large glass covered dial and decorations, it does just that.

In his April 1972 *NAWCC Bulletin* article on Morbier clocks, Lawrence A. Seymour explains that the large, weight driven Morbier style of clock began to decline after 1871, following the Franco Prussian War. Cheaper German goods entered France, and the market drifted towards smaller clocks. The French picture frame clocks were made in response to this trend. The article goes on to explain that many of the picture frame clocks had spring driven Morbier style movements, and we cover just such a clock in Chapter 5. Other picture frame clocks, such as ours, had conventional count wheel strike movements.

Japy Freres trademark from the back of the movement.

The marking on the dial of our clock indicates "J. ROI", the seller, and the location MEZIERES EN BRENNE.

The back of the movement bears a trademark, which Nicolas Thorpe in his book *The French Marble Clock* identifies as belonging to:

"Japy Freres et Cie., La Chaux de Fonds, Reg. September 1st 1887, Leipzig, Germany."

The movement has square, rather than round, plates.

To wind or set the clock,
unhook a latch under the
edge of the dial and lift it
from the bottom.

Disassembly

The first step with this clock was to hang it on the wall and see if it would run. That's a good way to get to know a "new" clock. Since it came from a dealer, I did not know whether the pendulum was the correct one for the clock. Numbers on the movement and the pendulum did not match, indicating the pendulum was a replacement. As it turned out, the clock ran overnight and kept close to correct time.

I removed the pendulum, then placed the clock flat on a workbench. Two of the movement bracket wood screws were too long, protruding through the case back. I'm happy to have noticed that, as the sharp steel points would have marked up almost any surface. A sheet of cardboard provided protection for the bench.

The hands came off first, after the taper pin was pushed out with pliers.

The next step was to locate the three dial pillar screws. These are very long, thin steel screws with brass heads. Oriented at three points underneath the dial, they are easily located by feel, unscrewed, and removed.

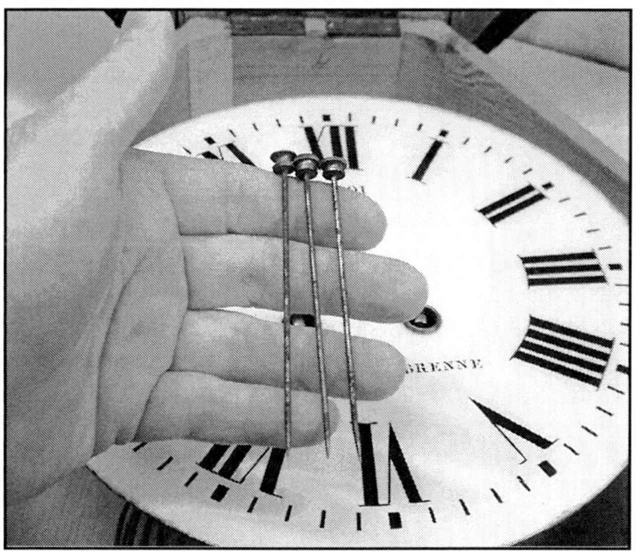

As I set aside the dial, I received my first look at the movement, the less common square French type. There is a lot of empty space around the movement! The pivoted wooden stick on the right side of the case, indicated by the white arrow, is a prop to hold the door open when the clock is on the wall.

I wanted to remove the three height adjusting screws on the side of the movement brackets. This would allow the movement to be taken out of the case. However, I could not budge the screw indicated by the arrow (right), at least not with the short screwdriver that would fit in the available space. So, I instead removed the three wood screws holding the bracketed movement to the case back. This did the trick, and I removed the movement as shown in the photo. This provided the added benefit of removing the extra-long screws that were waiting to scratch me or the bench top, sooner or later.

A quick look underneath the movement showed that the mainspring barrels had tabs bent in from the barrel wall, in place of mainspring hooks (rivets). I can't recall seeing these tabs in a French movement before. It seems a lower cost feature mostly associated with German clocks from the 20th century. At least I could see that the tabs appeared intact. Too often, tabs are broken off or split.

The hammer, shown near the gong in the middle photo, had no leather in the head. The head did not hit the gong efficiently, either. It appeared that the hammer shaft might have been cut short.

Then I saw the other set of screw holes for the gong base. Three empty screw holes are indicated by the black arrows. The repairer used one hole from that set of four and moved the gong base— or did he? Later work seemed to show that the gong's present location was the only one that would allow the gong wire to stay clear of the movement. Was the gong wire a replacement? It's a bit of a mystery. My solution was to plug and redrill the lower left screw hole, as it didn't quite match up with the gong base. I kept the base mounted as shown.

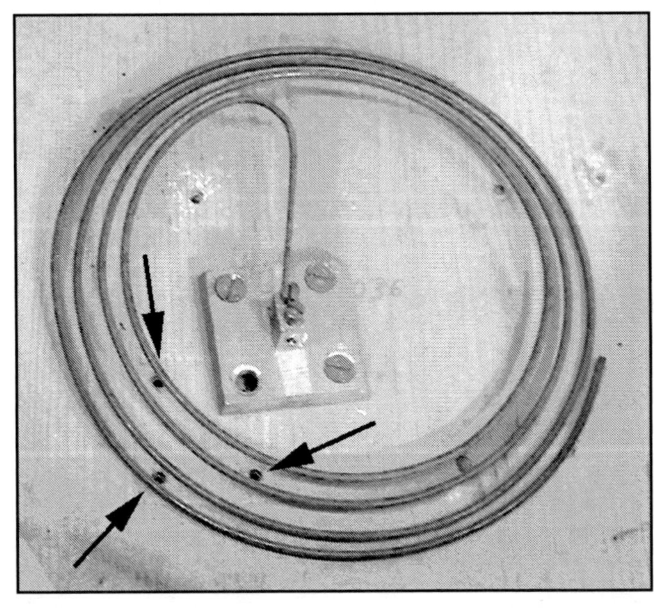

Let Down the Mainsprings

The first thing that must be done to a movement is to let down the mainsprings. A letdown key, available from any clock parts supplier, allows this to be accomplished as safely as possible. My left thumb pushes down on the tail of the click, raising it away from the clickwheel at the white arrow. The letdown key is allowed to spin, always controlled, in my right hand. After the first spring is let down, repeat the procedure on the other spring.

In this instance, I left the mounting brackets on the movement. They provided a stable support for the movement as I let down the springs.

Remove the Motion Work

Remove the screw holding the minute wheel cock, then take it off the movement, which allows the other motion work parts to be removed.

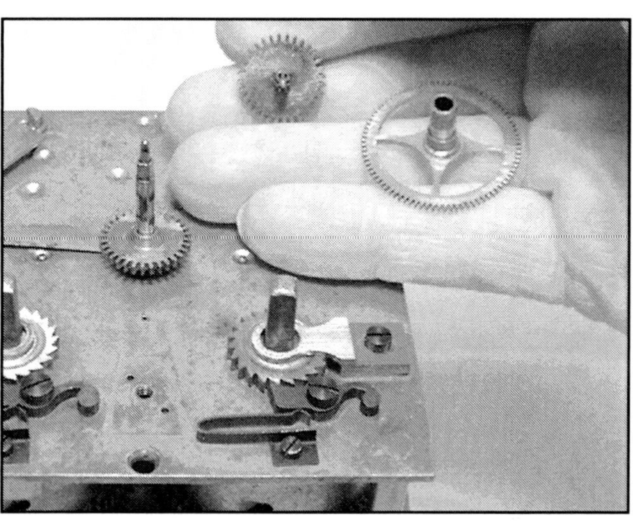

Remove the minute and hour wheels.

Pull off the cannon pinion with your fingers. If you found the minute hand tension was too loose, rest the slotted cannon pinion against a wood block and tap ever so slightly on each side with a small hammer. The *smallest* adjustment will firm up the hand tension. Do not risk crushing the cannon pinion tube with pliers! Push it back onto the center arbor to try the tension.

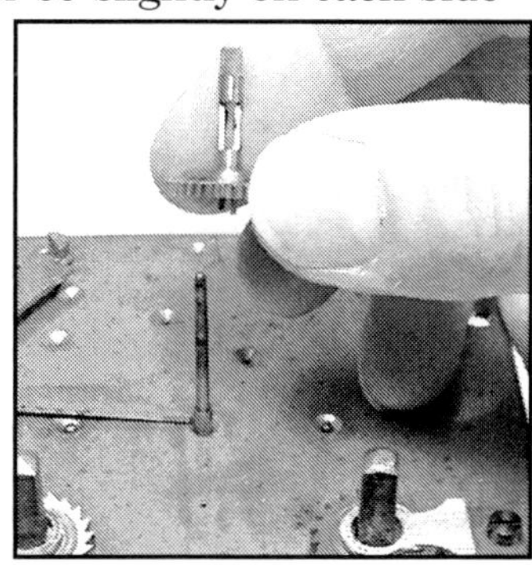

The lifting/warning lever assembly indicated by the arrow should be removed after the taper pin has been pressed out of the post.

Remove the click, so that the movement can be properly cleaned. The click and its special, shouldered screw are shown in the closeup view.

Then use a small screwdriver to remove the set screw that holds the clickspring in place—it's located beneath the click in this movement—and remove that, too. The time side clickspring is shown.

Remember that both mainsprings must be fully let down (previous page) before this step is started. Remove the set screw for the clickwheel cover. Take off the cover and the clickwheel.

At this point, all the parts have been removed from the front of the movement. The strike parts are shown placed to the left of the movement, and the time parts are shown to the right. The only distinguishing marks on the front of the movement are the numbers shown in the inset view.

The third and last bracket is shown being loosened for removal. There was no need to mark the locations of the brackets, as a previous repairer had "thoughtfully" marked two of them along with corresponding marks on the movement plates, next to each bracket. Mark, if you must, on the underside of a part. Do not mark the plates or visible parts of wheels. One could also say, "Don't mark at all". Just make notes and sketches or take photos.

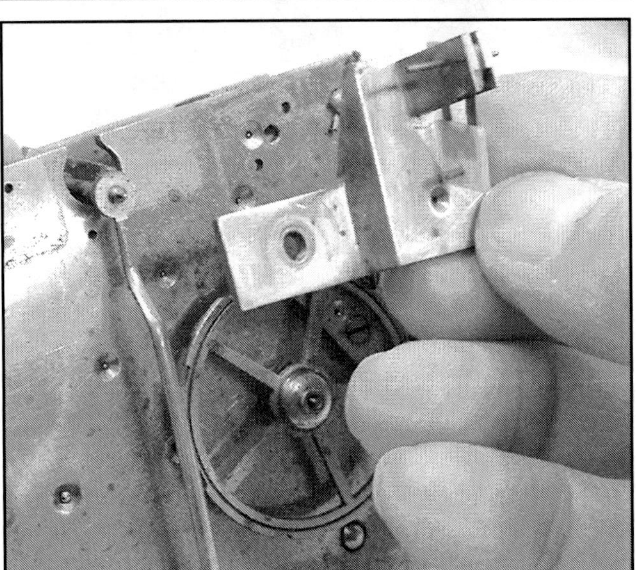

The back cock, which supports the pallet arbor and suspension spring, was removed next.

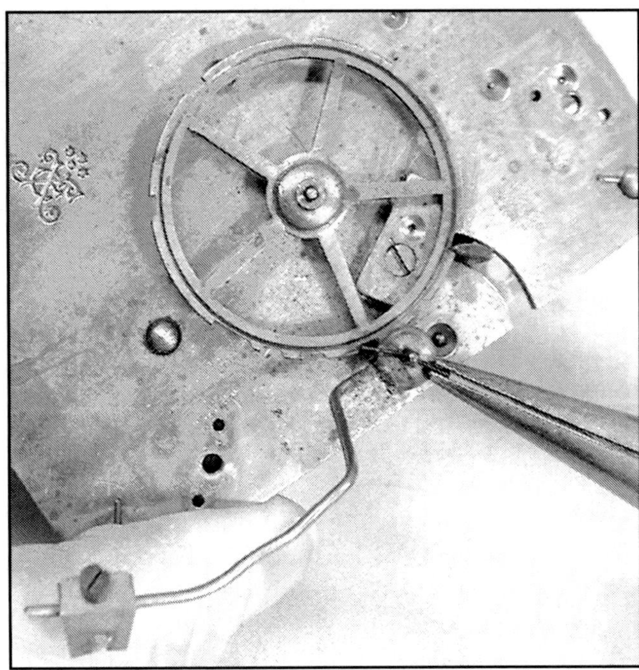

The taper pin holding the hammer on its arbor broke off a few seconds after this photo was taken. There was nothing left to do but break off the other side and file the pin flush with the arbor. This allowed the arbor to be removed from the plate in a later step. Getting the remains of the taper pin out of the arbor will also be done later.

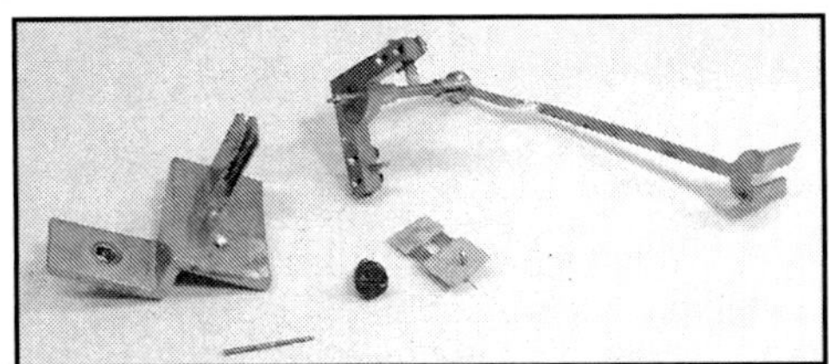

The back cock is shown at the left and the verge or pallet unit at the top. The suspension spring (center) was the wrong style but it worked perfectly well. I replaced it with a single bladed spring.

With the mainsprings let down, it is time to remove the pillar pins and separate the plates. The pins can be pressed or pulled out. Save the pins if you like, but they almost always need to be replaced. Use steel pins, not brass, for the pillars.

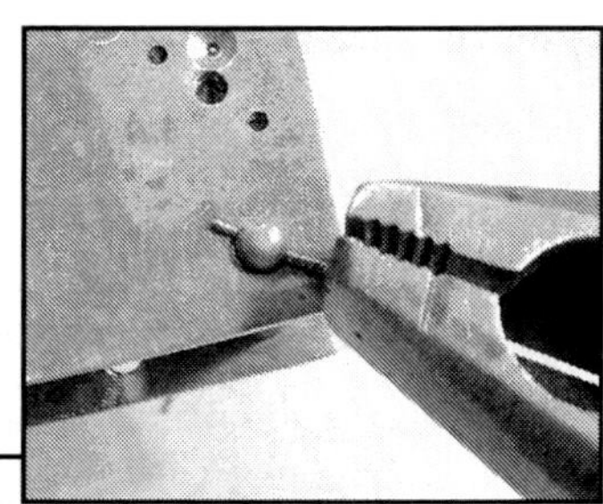

Separate the plates and avoid stressing or bending any of the fine pivots. It's nothing to be afraid of—just be careful!

You can see that the count wheel has not been removed from the back of the move- ment. It wouldn't come off, so it had to stay until a later step (see the next page).

Here is the disassembled movement.

I removed the screw holding the cock for the pinwheel rear pivot. Then I was able to remove the cock, even though it appears from this photo that the count wheel is in the way. However, the count wheel was free to move to the rear, as the plates had already been disassembled.

Removing the count wheel from the elongated square pivot can be difficult. The wheel hub has been driven onto a slightly tapered, elongated, square pivot. Add some corrosion and there is a strong bond to break. Pull out the taper pin first. Gently try to pry off the count wheel, before the plates are separated. This did not work on our featured clock, so I stopped. Additional, strong prying is not worth the risk of bending the pivot or the count wheel.

Here's how to remove a stubborn count wheel. With the plates sparated, tap gently with a small hammer on the end of the pivot, as shown in the photo. This utilizes the back plate as an anvil. The pivot is tapped through with very little effort, and there is no damage to the parts.

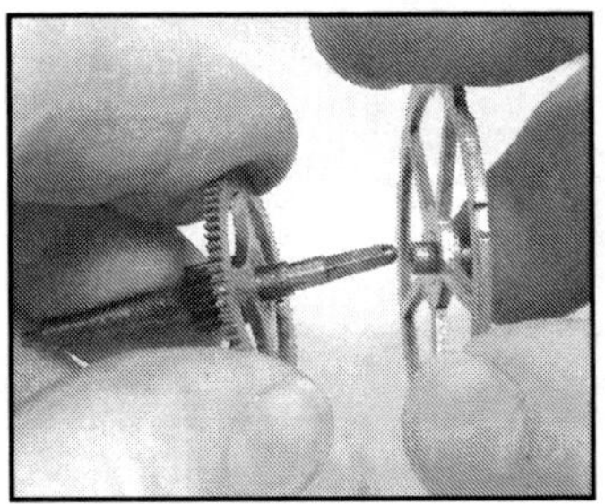

Before removal, it is a good idea to mark the position of the count wheel on the arbor. There are four possible positions. I have seen counting errors corrected by simply shifting the count wheel to one of the other three positions, presumably the original one. The original position is also likely to offer the best fit on the arbor. I failed to mark the position this time, but fortunately did not see any strike counting problems from it down the road. Maybe I was lucky and found the original orientation, or it may not have mattered in this movement.

The same removal procedure had to be used on the strike hammer. A few taps pushed the arbor through, separating the arbor from the hammer hub.

Here are the hammer arbor and the hammer. The hammer shaft is loose in the hub. We'll repair that later.

Mainsprings

Mainspring removal is next. Never skip this step!

It is sometimes possible to pop the barrel cover right off by tapping the rear of the arbor on a wood block. That didn't work this time, so I used a small screwdriver to pry the cover loose at the access hole. This will not damage the barrel if it is done carefully.

Removing Mainsprings by Hand

If you do not have a winder, remember that you will later have to wind the spring back in by hand. I find this to be very difficult. It requires considerable hand strength, and there is a risk of cutting oneself on the edges of a mainspring.

The procedure for removing a French mainspring without a winder is this: wind the arbor counterclockwise just enough to detach it from the spring. Pull out the arbor and set it aside. Wear safety goggles and heavy gloves. Grasp the inner coil with pliers and pull it free. With your hands, work the coils out of the barrel one at a time, in a controlled manner, until all the coils are free. Finally, push the outer mainspring hole end off the mainspring rivet. This hand procedure may distort the spring into a cone shape, and it may degrade the performance of the spring. I use a winder whenever possible.

For more information on handling mainsprings without a winder, see pages 22-24.

The barrel is placed in the winder clamp, with the barrel rivet oriented upward.

Wind the spring and insert the sleeve; let the spring down into the sleeve.

Turn the sleeve to unhook the mainspring outer end from the barrel rivet. Remove the sleeved spring from the barrel. Install the sleeved spring back on the winder.

The spring is wound slightly tighter, allowing the sleeve to be removed. Then the spring is let completely down, as shown here. It can now be taken off the winder.

The barrel, mainspring, cover, and arbor should be kept separate from the corresponding parts of the other gear train. Many movements are dot-punched to help, but our movement was not marked. I scribed a "T" for "time" in several places that would not show.

The mainsprings were found to be undamaged. That is, the hole ends were not torn and there was no rust on the coils.

The mainsprings measured 18mm wide x .015" thick x 58-1/2" long.

Cleaning the movement is the next step. Each pinion and wheel must be closely inspected for the presence of damage or hardened oils that can stop a clock. Rust is another enemy.

Install the arbors one or two at a time and look for worn pivot holes. Extra care is needed to place and finish bushings to give original appearance and performance.

One could repair two identical old clocks and find very different repairs to perform on each. One clock could have an escapement problem, and another could have damaged mainsprings or barrels. In our French wall clock movement, we had only two problems, aside from dirt and some rust on the steel parts. Both problems were related to the strike mechanism. One defect was a loose hammer shaft; it had to be corrected for the clock to strike the hours and half hours clearly on the gong. The other defect was less critical, because the clock would work whether or not it was corrected. This was an incorrectly installed cock supporting the rear pinwheel pivot.

Once the parts are clean, I will often work to install the mainsprings back in the barrels as my next step. Then I will proceed with other needed repairs.

We will begin now with the cleaned movement, ready for reassembly. The above mentioned repairs will be made along the way.

The mainsprings and barrels needed cleaning but presented no problems. I thought I might find cracks around the barrel hooks, as they are just tabs punched inward from the barrel wall. They were not damaged.

To begin, a cleaned spring is mounted
on the Keystone Mainspring Winder.

The spring needs to be wound tightly enough so that the diameter will fit in the winder sleeve. This is the same sleeve we used to bind our spring during removal (page 53). Lock the winder handle against the tool's locking rod while you install the sleeve. Then release the handle, controlling it carefully as you let down the spring into the sleeve.

The sleeved spring is now removed from the winder and inserted in the barrel. Hook the hole end of the spring onto the barrel hook. In this case our hook is a tab punched in from the barrel wall, not a steel rivet.

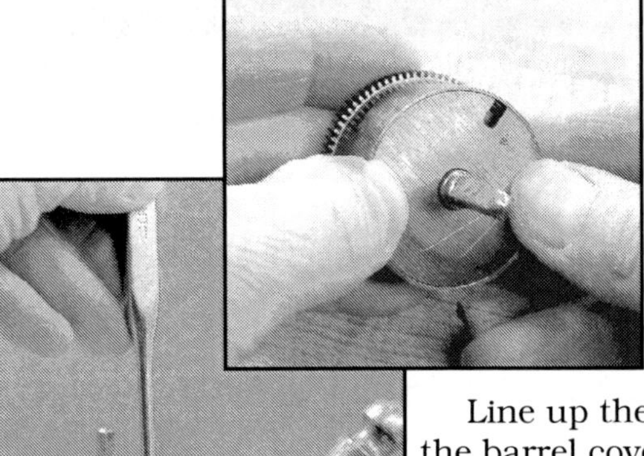

Line up the slot in the barrel cover with a dot-punch on the barrel or the toolmarks from previous removal. You may be able to press the cover on with your fingers.

The sleeved spring is gripped in the rubber lined jaws of the winder clamp. The barrel hook or tab should be oriented upward. Wind the spring to allow the sleeve to be removed. Then release the winder handle from the locking rod and slowly let down the spring within the barrel.

If the cover goes almost all the way on, you can gently squeeze it the rest of the way as shown on page 24. Instead, the cover can be gently tapped in place with a small hammer and flat punch, as shown above. If the cover does not pop into place right away, something is wrong. Stop and find out why the cover does not fit. Is it damaged? Is it the cover for the *other* barrel?

A Hammer Repair

Once I had the movement out of the case, I was able to see that the hammer shaft was very loose in the hub. Someone had soldered the shaft in place, the joint had corroded, and it had finally come loose from people twisting it. This is a very common problem in French striking clocks. I began thinking about the best way to make a new hammer shaft. It turned out it was not necessary.

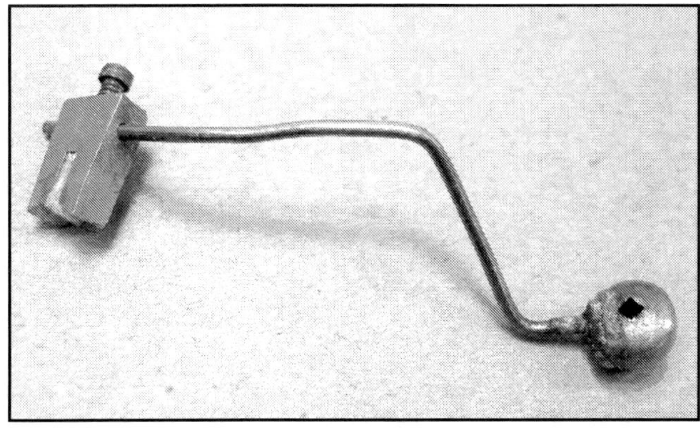

Almost by accident, I found the hammer shaft was *threaded* into the hub. I cleaned off all the solder and screwed the shaft tightly into the hub. Next, I carefully bent the shaft until it was close to the gong. I

separated the parts again, cleaned them with alcohol, and applied Loctite. After the parts were screwed together again, the joint was allowed to cure overnight.

When I tried the repaired hammer shaft hub on the hammer arbor, I remembered that the remains of the broken-off taper pin were still plugging the hole in the arbor. I used an anvil and a very small punch—just a bit of sharp scrap steel rod—and tapped out the pin easily. It is, of course, necessary to determine which side of the arbor carries the smaller end of the pin. Punch the pin from that direction. It may sometimes be necessary to drill out the remains of a pin, but in this case it was not.

The repaired hammer was checked by mocking it up in the movement with just the hammer arbor, pinwheel, and locking wheel installed between the plates. After replacing the leather in the hammer head, I rotated the wheels and was able to hear the gong sound clearly. The hammer was repaired.

Reassembly

Load the parts into the front plate and arrange them as shown. Strike train adjustments can be made after the back plate has been added.

After you place the back plate lightly in position, ease the pivots into the holes. There should not be a great fear of bending or breaking a pivot, as long as you follow these rules:

- Use tweezers.
- Never force a pivot.
- Always look for the next pivot that "wants" to go into place, and fit that one next.

Look for three conditions as you check your assembly of the strike train:

- The lock pin should rest against the locking lever.
- The hammer tail should be free of the pinwheel pins.
- The warning pin should be located 1/4 to 1/2 turn from the warning lever.

At the beginning of the chapter, I said that there were two main problem areas in this clock, one being the hammer that we have already repaired. The other problem is the bar shaped cock that carries the rear pivot for the pinwheel arbor. Incidentally, most French movements have three of these removable pivot supports, but our movement has only one.

The cocks have two steady pins to precisely locate their positions on the clock plate. I saw that both pins had been clipped off this cock and filed flush. I replaced both pins and, in the process, I found out why they had been removed. The two steady pins and the set screw could not all fit at the same time: the position was a little off. It is entirely possible that the cock was not original to this movement. I might add that there did not seem to be a problem in running the strike train with no steady pins at all. I compromised by replacing one pin and leaving the other out. Having one pin and the set screw to spot the position of the cock worked better than what was there before. There was no question of the cock being located incorrectly or varying each time it was removed and replaced.

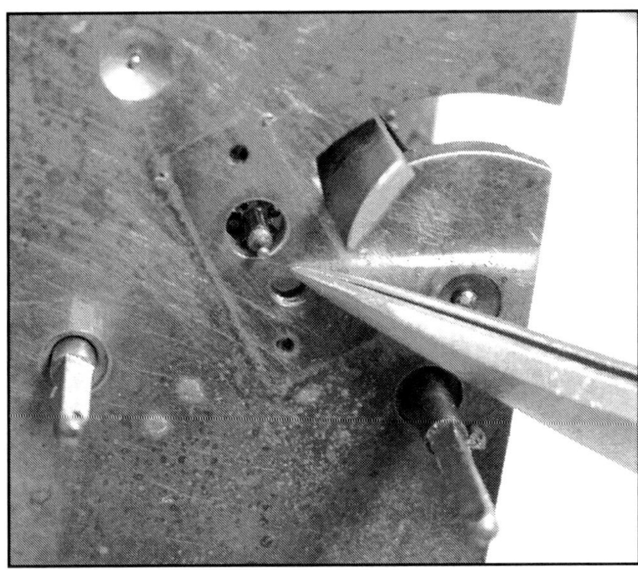

The tweezers point to the rear pivot of the pinwheel arbor, visible in a hole in the rear plate. The cock has been removed at this stage.

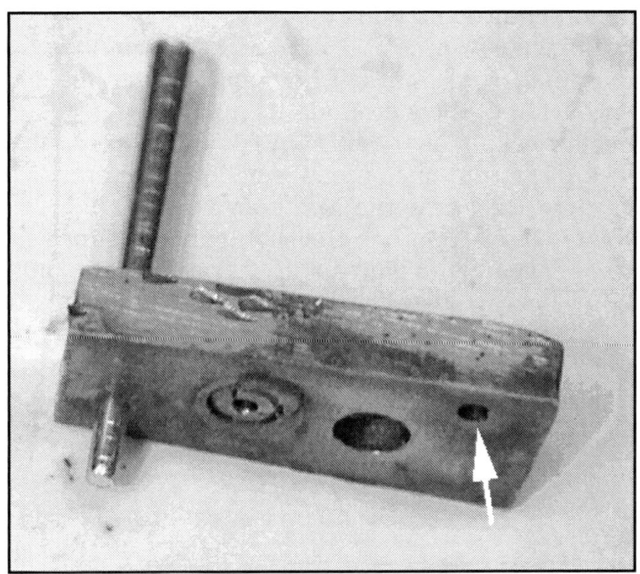

The cock is shown with one steady pin in the process of being replaced with a brass taper pin. The thicker, top portion of the pin will be clipped off and filed/polished flush with the top surface of the cock. The other steady pin hole, which I decided not to use, is indicated by the arrow.

The cock is shown being tried on the back plate with the set screw and one steady pin (not yet finished off).

The count wheel can go on the square arbor four ways, but an impression left by the taper pin on the hub of the count wheel will help you to eliminate two of them as wrong. The count lever will probably not enter the slots at the right places if the wheel is put on one of the two wrong

ways. The other two ways will be correct or nearly so. I found that this clock struck properly for a week and then began to get ahead in the count. This happened be- cause the locking action ending the half hour strike was marginal. The count lever stopped against the end "wall" of the slot, and this made the lever begin to climb as it locked. A few times a day, the lock pin would miss the locking lever, and the clock would strike the next hour immediately. I was able to straighten the slightly bent count lever assembly to gain some clearance at the half hour locking position. If more of an adjustment was needed, the count wheel would have to be loosened on its hub so that it could be turned to a different spot.

Install the cannon pinion on the center arbor. Follow with the clickwheels, the clicks, and the covers for the clickwheels. The clicksprings are next.

Add the the minute wheel, the hour wheel (the large wheel in the photo), and the minute wheel cock. The way these gears mesh is not critical (but it is critical in a rack strike movement).

Continuing with the reassembly, oil the pivot hole for the minute wheel, as it is hard to reach after the wheel is in place.

This view of the completed front of the movement shows all the parts in place.

After the verge or pallet unit has its front pivot placed in the front plate, the back cock can be added and held with its screw.

A front view of the completed movement

A replacement single bladed suspension spring is shown being inserted in the jaws of the suspension block. This block is not adjustable for length, so all rate adjustments are done with the pendulum. The top block of the suspension spring should fit freely in the jaws but not be loose. Fasten with a taper pin. File the ends of the pin smooth.

A rear view of the completed movement

5

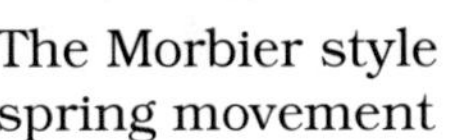

The Morbier style
spring movement

MORBIER SPRING WALL CLOCK

So-called French picture frame clocks were made with two different types of movements. Chapter 4 covered the disassembly and repair of one of these clocks with a rectangular, count wheel movement. The overall design has some features in common with the round plated French mantel movements with a similar strike mechanism.

Our subject in this chapter has the other style movement, variously called a Morbier after the French village, or after Morez, a larger town in the region. Steel bars carrying the brass-bushed pivot holes are mounted in an iron cage (see the photo at the top right of this page). The movement is similar to the weight driven Morbier wall clocks with brass dial surrounds and large, hammered brass pendulums. The main difference in the featured movement is that two mainsprings provide power in place of weights.

The movement has a distinctive ladder rack strike that sounds the hour, then repeats the strike a few minutes later. The movement strikes on a wire gong.

The case looks similar to the Chapter 4 clock. An ebonized plaster-over-wood frame borders the veneered front with its various wood decorations, brass work, and mother-of-pearl inlay. The dial of this particular clock is painted metal, rather than reverse painted glass as in the other clock.

Our Morbier picture frame clock
is 24-1/2" tall x 19" wide.

Remove the taper pin holding the minute hand in place.

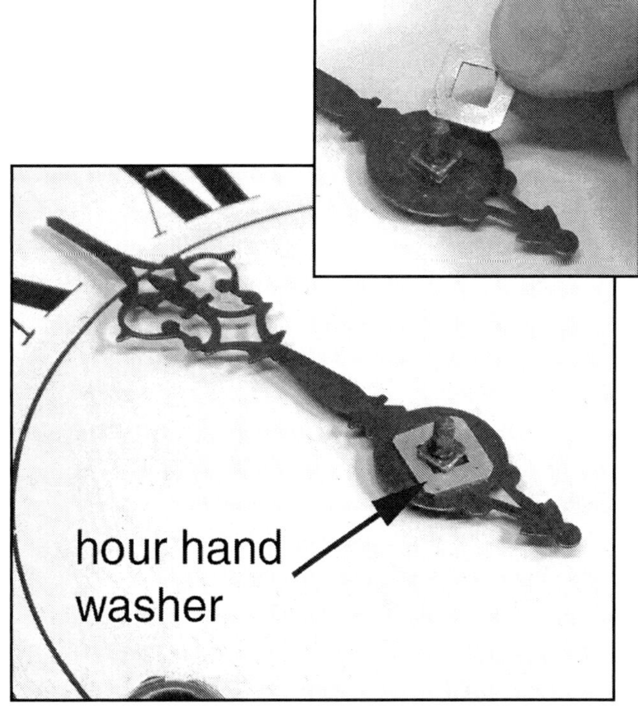

Disassembly

Lay the case on a bench and support it on wood blocks to prevent scratching the table top. With a much thicker block at the bottom, the case will lean back, and the large door will stay safely in the upright position. Check this carefully for your clock. The door must not be allowed to fall!

It is necessary to remove the door to clean the inside surface of the glass, replace the glass, or clean the veneered surface. Cleaning the inside glass surface and applying some wood polish to the inlaid veneer will greatly improve the appearance of the clock. The door is easily removed by taking off the hinges. To separate the outer door frame from the veneered panel, remove the wood screws which hold them together. The white arrows in the photo show the screw locations, except for one screw that is hidden from view behind the hinges. With the door removed, all screws are accessible.

The wood backing for the veneer warps badly in some clocks, sometimes splitting the veneer. You may find extra screws were added to pull in or arrest the warpage. This kind of damage is something to look for when considering a clock for purchase or repair.

In our featured clock, the hour hand washer was missing, and a piece of soft wire had been crudely wrapped around the arbor in its place. I made the washer shown here, starting with a short length of 1/2" diameter brass rod mounted in the lathe. I drilled it, parted it off .025" thick, then filed the square portion to fit the clock. The hour hand was put in place, then the washer was added and turned about 45° in a groove. This locked the hour hand in position. The thickness of the washer and dimensions of the squared hole will not be the same for each clock.

This clock has three dial pins. Each one is a long piece of unhardened steel wire, tapered at one end to fit through the hole in a dial pillar. The other end is curled around so it can be gripped with the fingers.

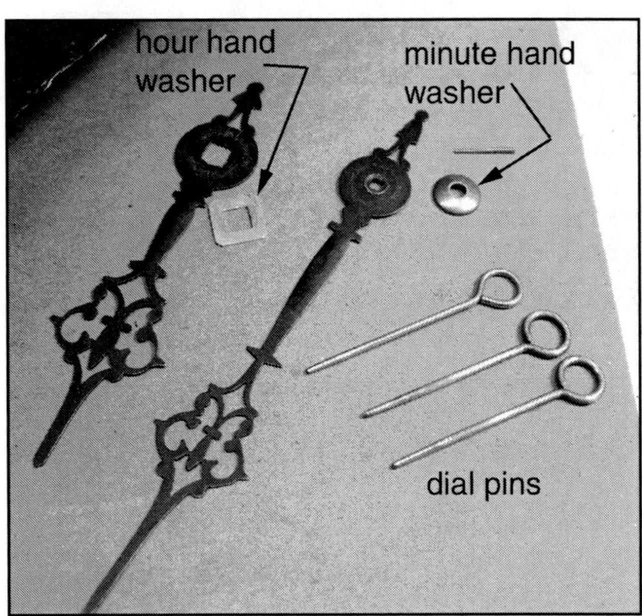

The photo shows the small parts removed so far. One of the three dial pins was missing; I made the top one from 1/16" diameter steel wire. The minute hand washer is also a replacement, this one found in a parts box.

After the dial pins are pulled out, remove the dial and set it aside in a safe place. This particular dial was carefully checked to make sure the numerals were not created with a waterbased ink or touched up with such an ink. Once satisfied with that, I was able to clean the dial with a soft cloth and a mild soap solution. Not all marks would come off, but the overall condition of the dial was very good.

The heavy movement is fastened to the case back with brackets. A matched set of machine screws (see arrows) and flat nuts on the back of the case hold the brackets securely. These fasteners may be original.

These views show the heavy Morbier movement removed from the case. The mainsprings were found partially wound up, but the clock would not run or strike because it was extremely dirty.

On the back of the movement, I found a broken suspension spring. I already knew the pendulum was missing; the dealer provided a French pendulum that I could modify to fit.

The hammer is a bit unusual: the brass head is slotted. The leather stretches around a wooden plug which was removed for this photo.

Before beginning work, always use a letdown key to release the power from the mainsprings (in this clock, there are two mainsprings).

The mainsprings must be let down before you perform this step!

The clickwheel cover is a brass strip that retains both clickwheels. Remove the set screw and take off the cover, taking care not to lose the spacer bushing underneath it.

The hour wheel and snail unit is simply pulled off the center arbor.

The motion work wheel assembly drives the minute hand with its wheel, and it drives the hour hand with its pinion. The leaf spring indicated by the arrow provides tension; it enables the motion work wheel to move the hands, yet it allows the minute hand to be turned to set the time.

Remove the taper pin (inset view) and then pull off the wheel assembly.

The top movement bracket
is shown being removed.

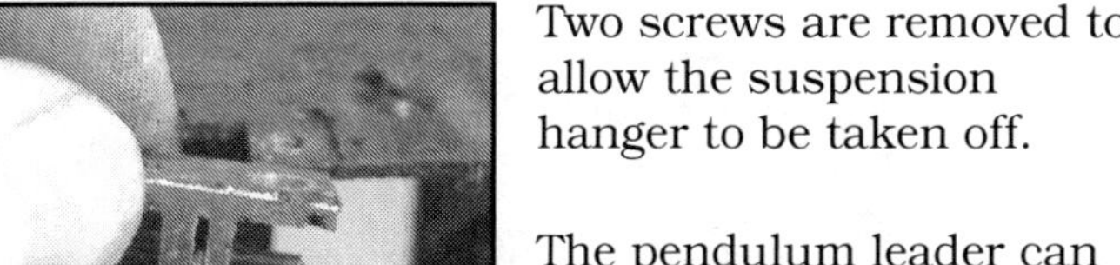

Two screws are removed to
allow the suspension
hanger to be taken off.

The pendulum leader can
now be unhooked from the
crutch (at arrow, below)
and removed.

A taper pin is withdrawn at
the back of the movement...

...permitting the center
arbor to be pulled out from
the front of the movement.
The center
arbor is carried
in the center
frame member
of the move-
ment "cage",
not in either of
the vertical
steel strips that
carry the time
and strike
pivots.

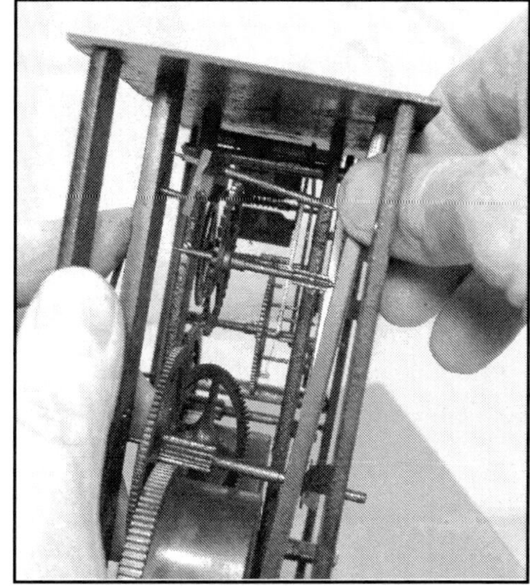

Take out the four screws from the
top of the movement frame or cage
to release the steel strips that
support the time and strike trains.
As shown here, pull the steel strips
outward at the top to angle them
out. Some of the arbors will start to
fall out as each strip is angled out.
The bottoms of the strips are mor-
tised into the floor of the clock;
that's why removal begins at the top.

The time train
components

This is a side-view of the strike parts.

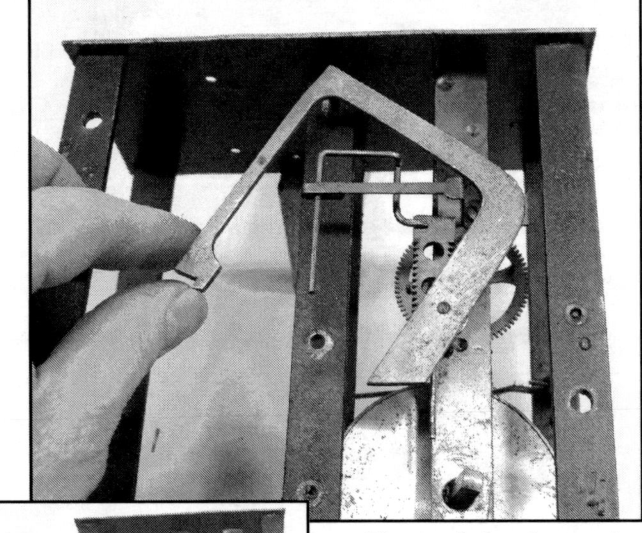

The rack hook must be removed from this movement before the strike barrel will come out of its cage "prison". Unpin the rack hook arbor and pull it out. Some angling and fussing will be needed.

The arrow points to a replacement stop pin someone installed for the hammer arbor. The extra top portion can be dressed down or re-moved.

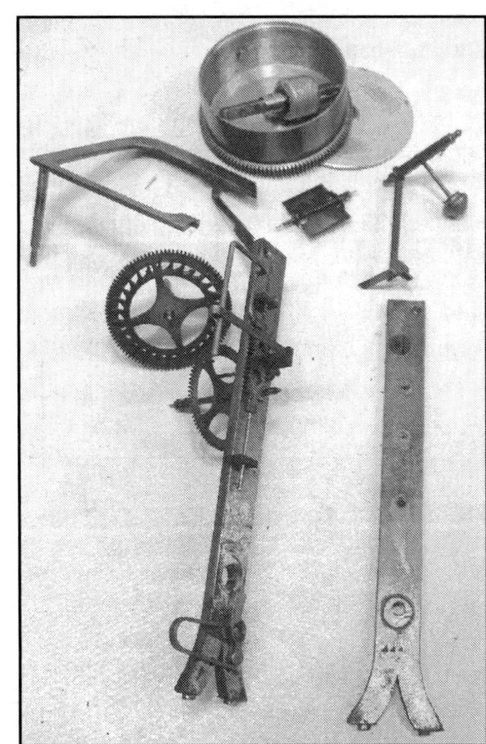

The strike components are shown after being extracted from the move-ment. It was difficult to remove these parts; will it be just as hard to get them back in place?

This reference view may come in handy later, when it's time to reassemble the movement.

The barrel is quite large at about 2-5/8" outer diameter. It was fortunate that this would fit in the Keystone Mainspring Winder clamp. The barrel wheel is about 2.9" in diameter.

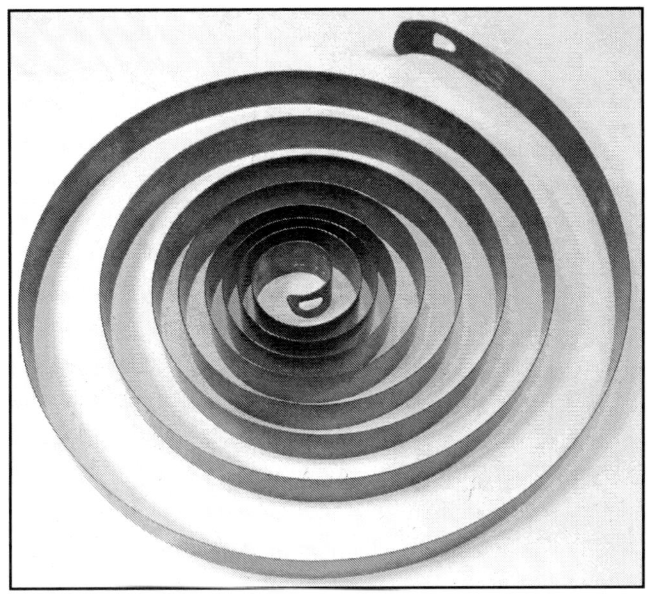

The two mainsprings are the same size. Fortunately, both were free of rust or cracks. Here are the dimensions:

    length: 113"
    width: 21mm or .827"
    thickness: .017"

Here is a close-up of the trademark stamped into the time barrel cover.

The last part of the disassembly process requires removal of the gathering pallet. The pallet, indicated in these photos, is a press-on taper fit. Do not twist it off, which is risky for the elongated strike pivot. Lay the steel movement strip on a split stake or even across two wood blocks. Make sure the lower end of the strike arbor is free. Tap the end of the pivot downward to separate the gathering pallet from the arbor.

Reassembly

In a posted movement such as our example in this chapter, the top and bottom plates are part of a rigid cage supported by riveted posts. In comparison to a standard plated movement, the Morbier spring driven mechanism offers a different experience to the person who assembles it. The time and strike trains are separate, and each can be assembled and serviced independently. The frame, or cage, as shown in the photo, is fastened together permanently. Earlier in this chapter, we saw that each gear train is supported in a pair of bars slotted into the bottom plate. The bars (*not* shown in the photo) are retained at the top by set screws through the top plate.

The strike gear train, minus the spring barrel, is shown flat on the bench in this photo. There are four arbors above the barrel. The largest of these is the pinwheel, followed by the gathering or locking wheel, then a plain fourth wheel, and finally the fly.

The gathering pallet is a brass piece that has two equal arms on it. It has a tapered center hole and is pressed onto the arbor with the fingers.

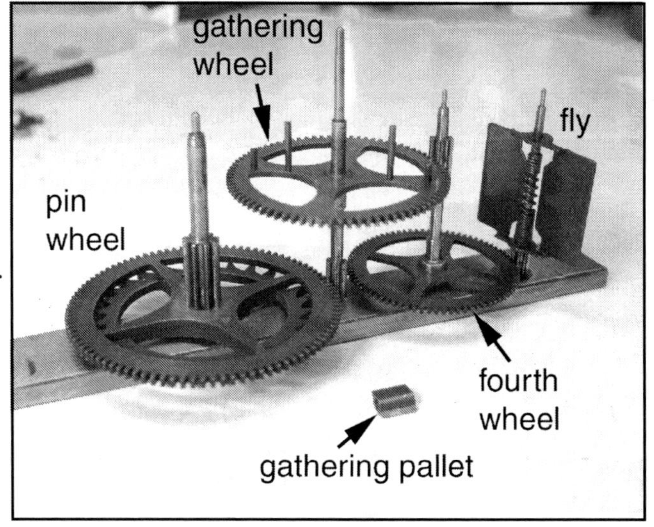

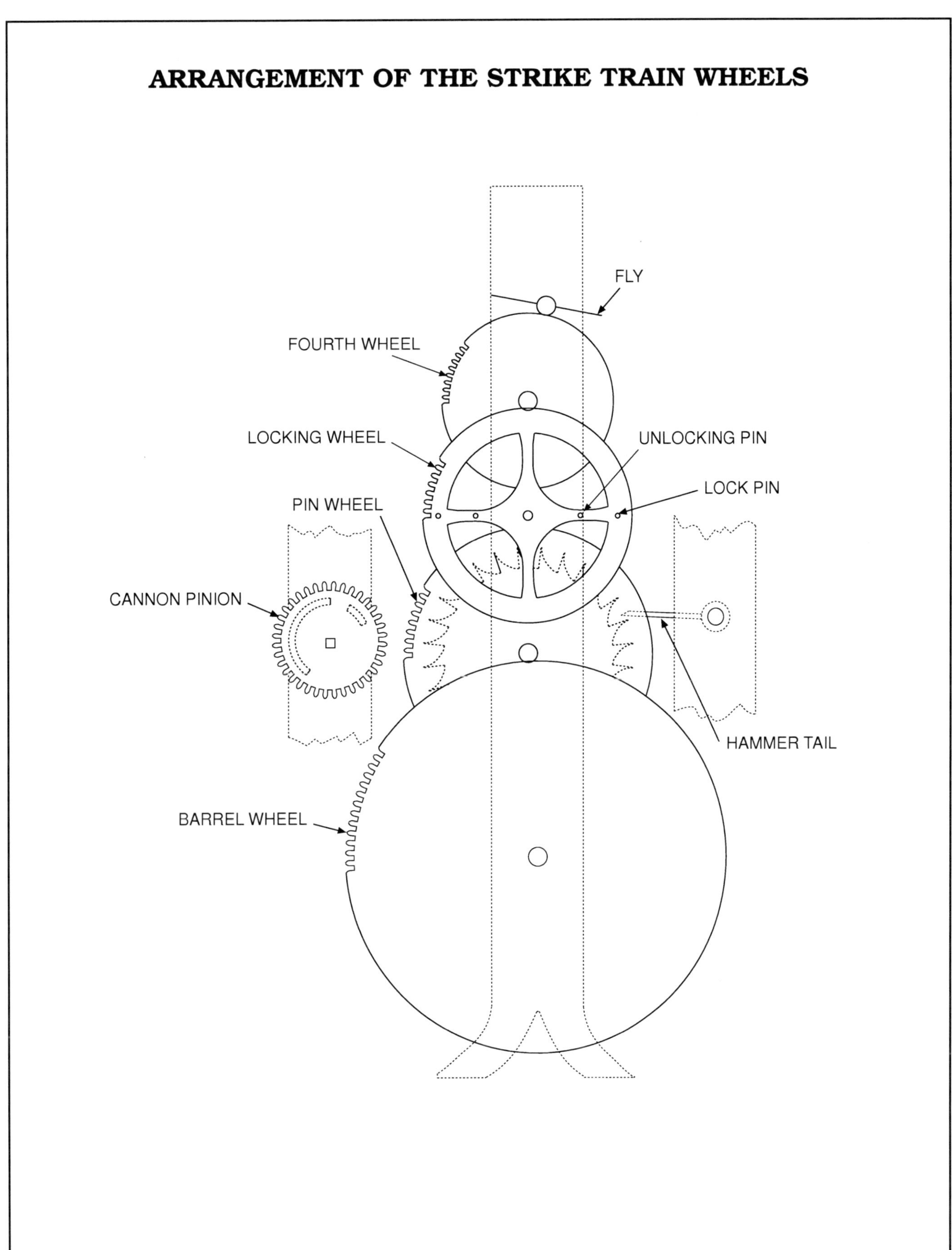

ARRANGEMENT OF THE STRIKE TRAIN WHEELS
FLY
FOURTH WHEEL
LOCKING WHEEL
UNLOCKING PIN
LOCK PIN
PIN WHEEL
CANNON PINION
HAMMER TAIL
BARREL WHEEL

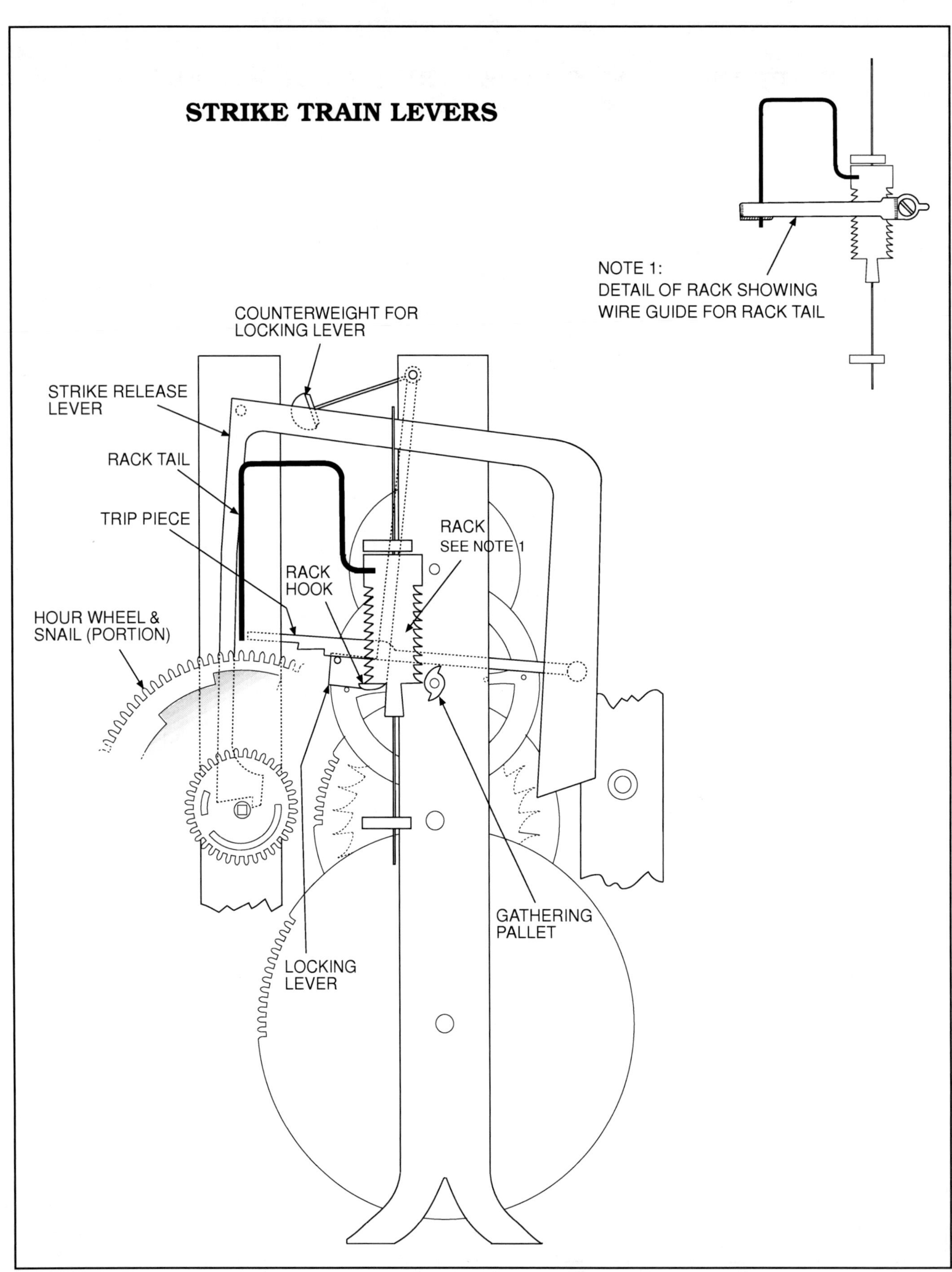
STRIKE TRAIN LEVERS
NOTE 1:
DETAIL OF RACK SHOWING
WIRE GUIDE FOR RACK TAIL
COUNTERWEIGHT FOR
LOCKING LEVER
STRIKE RELEASE
LEVER
RACK TAIL
TRIP PIECE
RACK
HOOK
HOUR WHEEL &
SNAIL (PORTION)
RACK
SEE NOTE 1
GATHERING
PALLET
LOCKING
LEVER

After installing the rear strike train pivot bar, fasten it at the top with a screw. I had taken 15 minutes out to lathe-polish the set of four screws and then blue them with a torch. This is worthwhile, as it makes the movement more presentable.

For a moment, it seems that the strike barrel won't fit between the vertical posts of the frame. Remember, the frame posts cannot be moved. But the barrel can be angled into position! (After all, it did come out.)

With the rear bar and the strike barrel in place, I placed the strike wheels in the clock. I soon realized there was a problem. It didn't seem possible to insert the front bar in the bottom plate of the frame and angle it backward. The long winding arbor was in the way.

So, right after I took this photo, I took everything out again except the barrel. I installed the front bar first, then the wheels, and then added the rear bar. This worked much better.

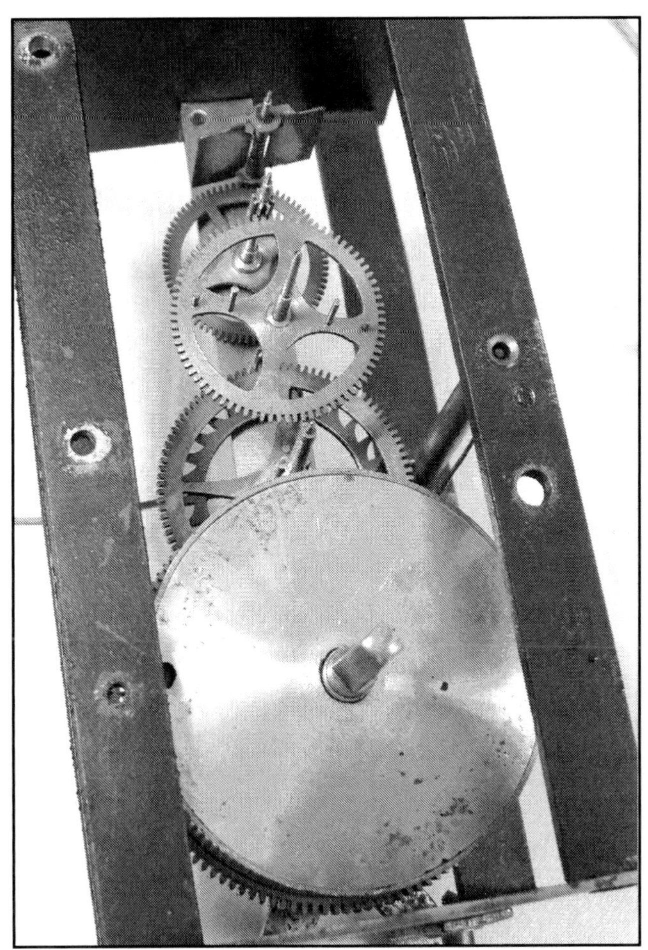

This isn't quite the way I ended up assembling the strike train! But these two photos show the counterweighted locking lever and the way it fits into the movement.

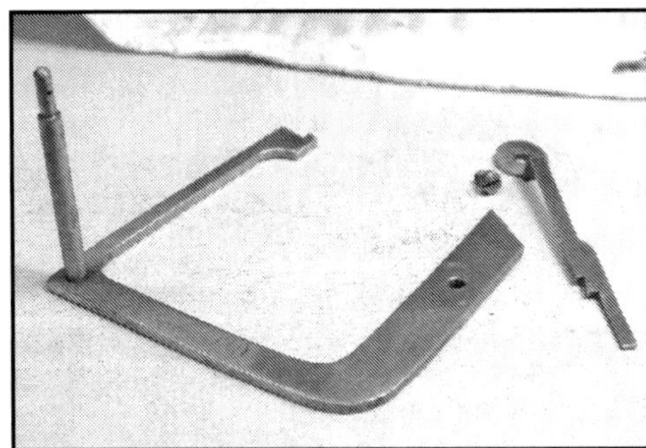

1. With the wheels in place and both bars installed, the strike release lever was the next part to install, but it would not fit in. I unscrewed and removed the trip piece, as shown in the photo. It was still necessary to take out the top screws and angle both bars back a bit at the top. Then the strike release lever fit in.

3. The last step was to slide in the trip piece and fasten it with the screw. The trip piece must swing freely on the screw.

2. The strike release lever (an upside down "U") is shown in place. The arrow points to the screw hole for the trip piece, which was removed.

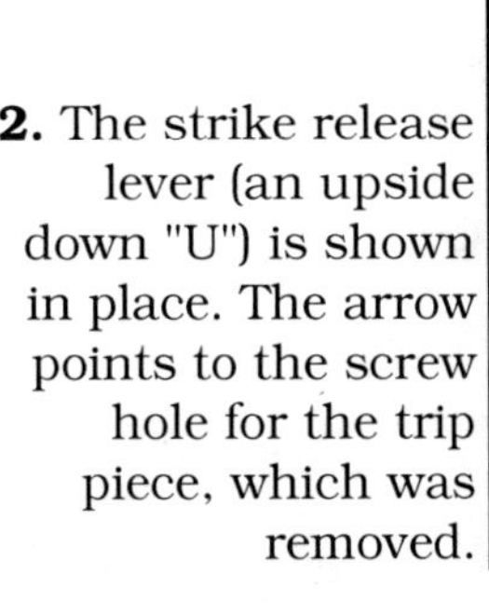

Tape the hammer out of the way.

The Morbier strike train does not utilize a "warning". The train starts quite suddenly, as the trip piece moves to the side to unlock the train. As a result, there is only one strike assembly adjustment to make: the hammer tail must rest clear of the pinwheel.

It is unlikely that the train will be assembled correctly the first time. To make the adjustment, proceed as follows:

1. Make sure both mainsprings are let down.
2. Tape the hammer out of the way, as shown in the photo.
3. Take out the set screw at the top of the rear strike pivot bar.
4. Move the bar rearward at the top.
5. Ease the rear gathering arbor pivot out of its hole.
6. Move the pinwheel to a new meshing point.
7. Verify that the hammer is at rest with the train locked.

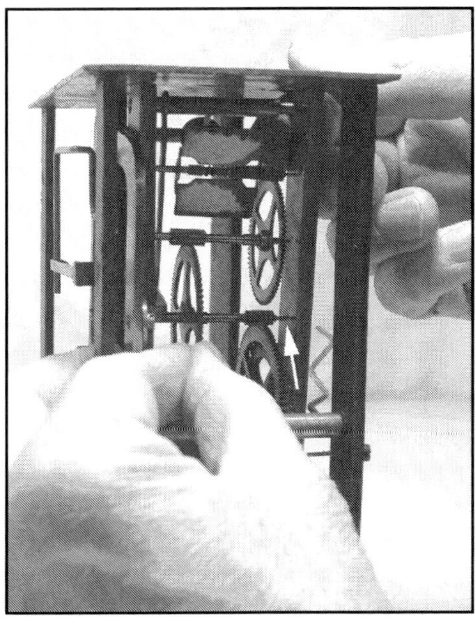

Ease the pivot out of the hole (see arrow).

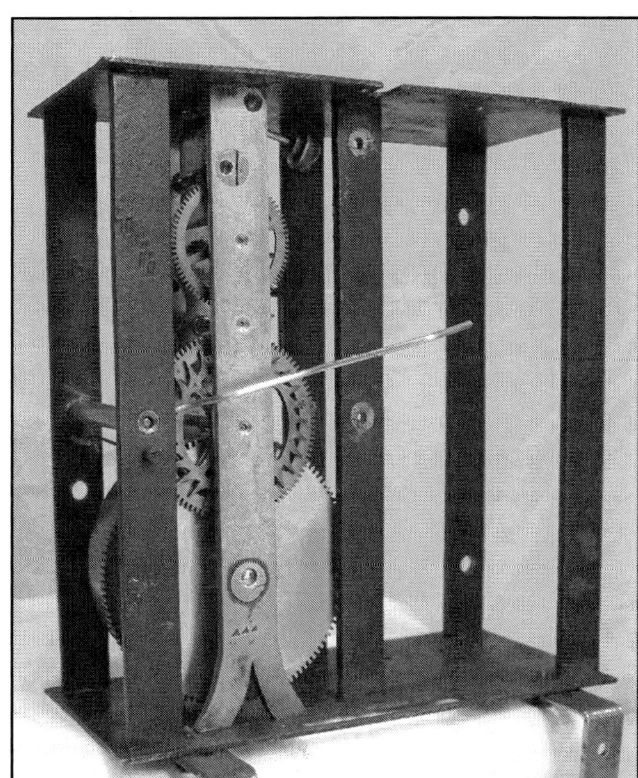

This is a view of the assembled strike train.

The arrow points to the hammer tail which is clear of the arms of the pinwheel. The gears must be meshed "just so" to make this happen.

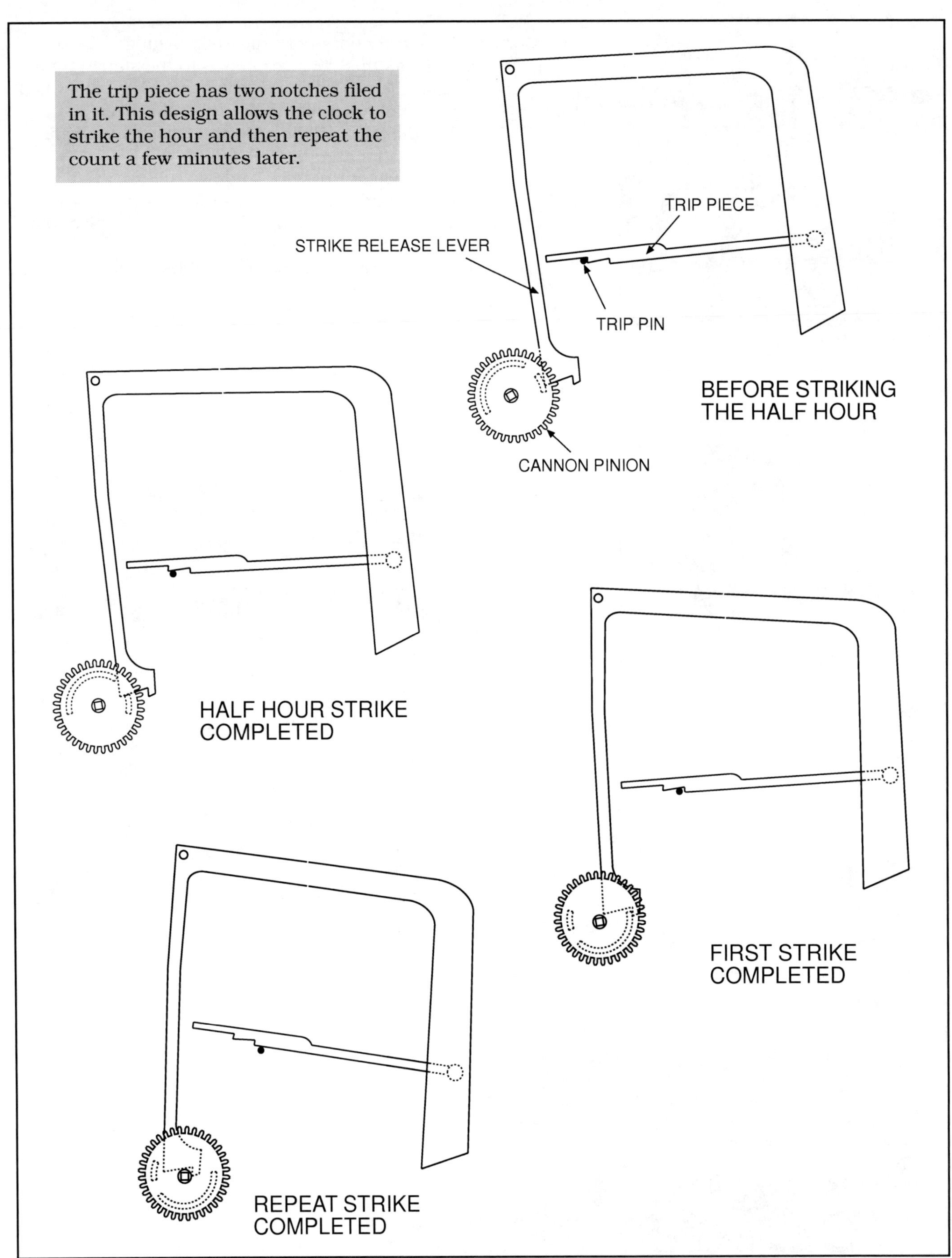

The trip piece has two notches filed in it. This design allows the clock to strike the hour and then repeat the count a few minutes later.
TRIP PIECE
STRIKE RELEASE LEVER
TRIP PIN
BEFORE STRIKING THE HALF HOUR
CANNON PINION
HALF HOUR STRIKE COMPLETED
FIRST STRIKE COMPLETED
REPEAT STRIKE COMPLETED

The time train is much easier to assemble than the strike. Install the rear bar. Add the mainspring barrel, then notch the front bar into the bottom plate of the frame. Leave the front bar angled out at the top...

..and install the rest of the time train wheels. Even the second arbor, with its long front pivot, fits in easily.

The pallet unit can be angled in through the hole in the rear bar. This needs to be done before the front bar is moved into its final position. Finish up by easing all the time train pivots into their holes in the front bar.

Install the set screw at the top of the front bar.

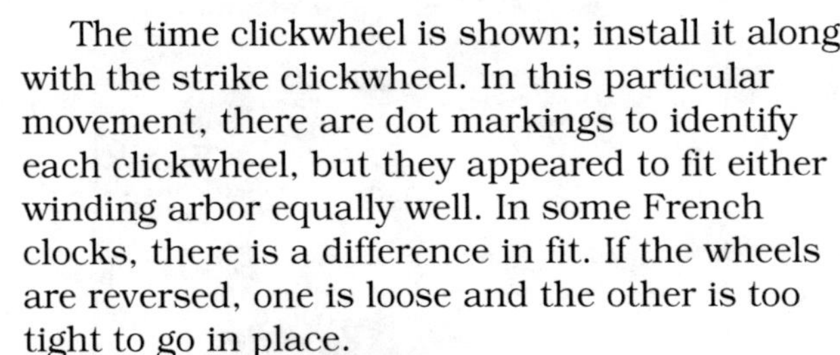

The time clickwheel is shown; install it along with the strike clickwheel. In this particular movement, there are dot markings to identify each clickwheel, but they appeared to fit either winding arbor equally well. In some French clocks, there is a difference in fit. If the wheels are reversed, one is loose and the other is too tight to go in place.

The next part is the single cover for both clickwheels. It is supported by a central screw. Note the spacer bushing underneath the cover which allows the screw to be tightened without binding the clickwheels. There did not seem to be a required left-to-right orientation of the cover, but there is an inside-outside one! The set screw has a chamfered shoulder that fits properly with the cover right side out.

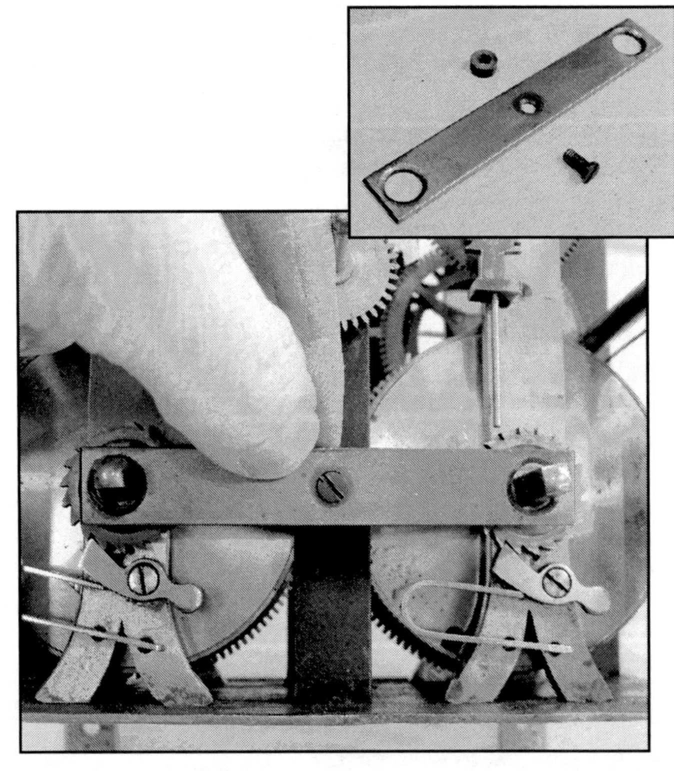

The hand tension spring fits onto the elongated front pivot of the time second arbor. Each end of the spring should be lubricated with clock grease (the toothpick points to one of the ends). These surfaces will rub against the underside of the minute wheel to provide hand tension.

Push on the minute wheel (left, in the photo above) and then fit the hour wheel and snail. Turn the hands to make the rack drop for the hour. Mesh the hour wheel so the rack tail will fall upon the 12 o'clock segment of the snail. Then check the 1 o'clock position (right).

When the snail is set correctly, install a washer and a taper pin to hold the minute wheel in place. It should be necessary to push on the wheel to obtain clearance for the taper pin. Try the minute hand to check for acceptable hand tension.

Here are the components of the suspension spring assembly. An ordinary suspension spring for an older Hermle movement fit into the slot after some filing and polishing. It may not be strictly correct, but it functions well. The pin was driven out of the bottom part of the suspension spring, and both holes were enlarged to fit the taper pins.

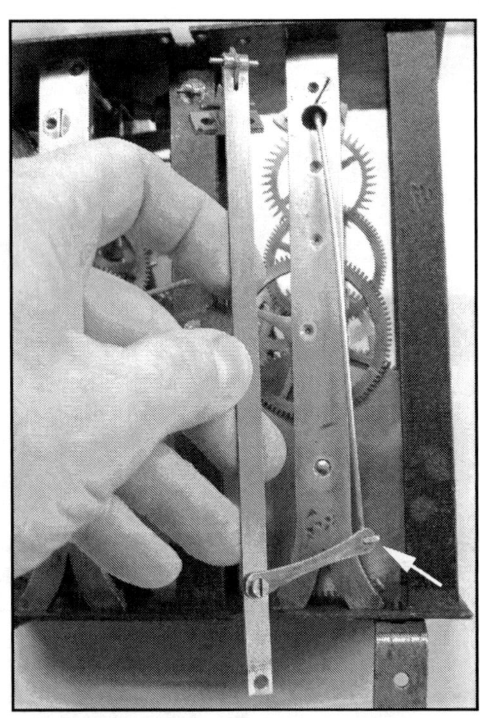

Hook the pendulum hanger's connecting piece to the end of the verge as shown at the arrow.

The suspension hanger and suspension spring are oriented in a way that is different from most other clocks.

Here is a close-up view of the connecting piece in position to run the movement.

The original pendulum was missing from the clock. This is the one the dealer supplied. I shortened it considerably to fit in the case and run the movement at the correct rate. The brass top hook received a no-frills modification to enable it to go through the hole in the bottom of the pendulum hanger; I simply squeezed the double hooks together in a vise and filed them thinner. As modified, the pendulum is 4" long overall and weighs in at 9 ounces. The brass faced lead bob is 2-1/4" in diameter.

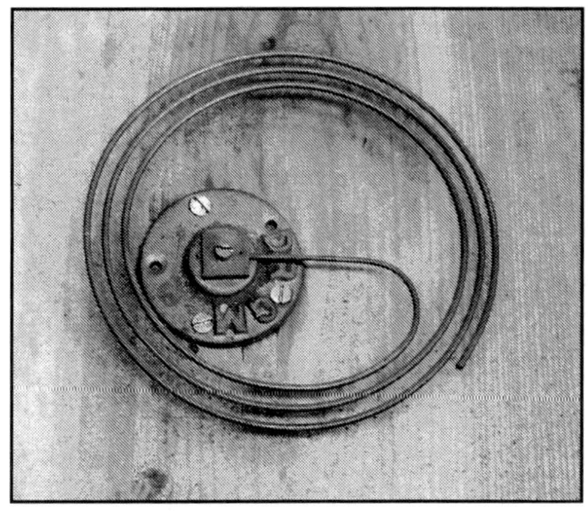

Unfortunately, the gong is not original, either. Bearing the letters D.R.G.M., signifying a German patent, the gong makes the correct sound for this clock. Adjusting the hammer to strike properly is done by bending the hammer shaft, not the gong.

One of the final tasks in setting up the clock is to check for even pendulum beat. Adjusting it is awkward, since the crutch is on the back of the movement and it is made of strong steel that takes some effort to bend. Install the movement and hang the clock straight on the wall. Hang the pendulum and listen to the beat. Determine the way the crutch must be bent. Take down the clock and remove the movement from the case. Grasp the crutch at the top arrow and bend it at the bottom arrow. Then set up the clock and check the beat again.

6

A SILK THREAD SUSPENSION

The older 19th century French movements were fitted with a pendulum suspension of ancient design that originated with Huygens in the 17th century. Instead of a geared regulator supporting a brass-and-steel suspension spring, the early setup is basically just a loop of silk thread attached to two arbors. The pendulum swings back and forth on this loop. The loop is adjustable for length, since it is wrapped around the upper arbor. This arbor pierces the dial, where it can be turned with a small key.

The trademark of the movement in our featured clock bears the name of Pons, an early 19th century maker.

The featured clock is a beautiful brass piece that stands 16-1/4" high and is 12" wide.

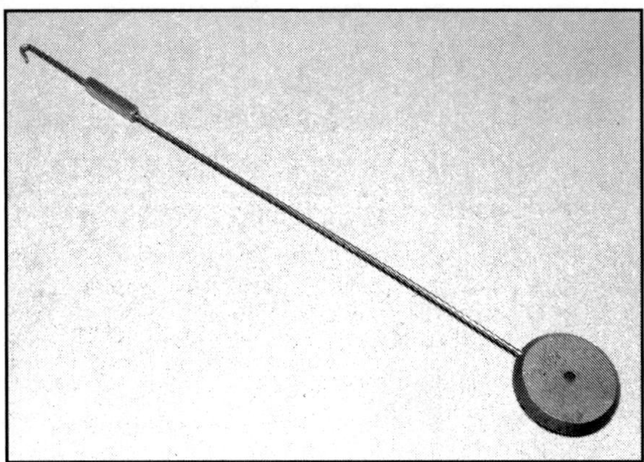

The pendulum features a side oriented hook at the top, a brass block to fit the crutch loop, and a small, non-adjustable bob.

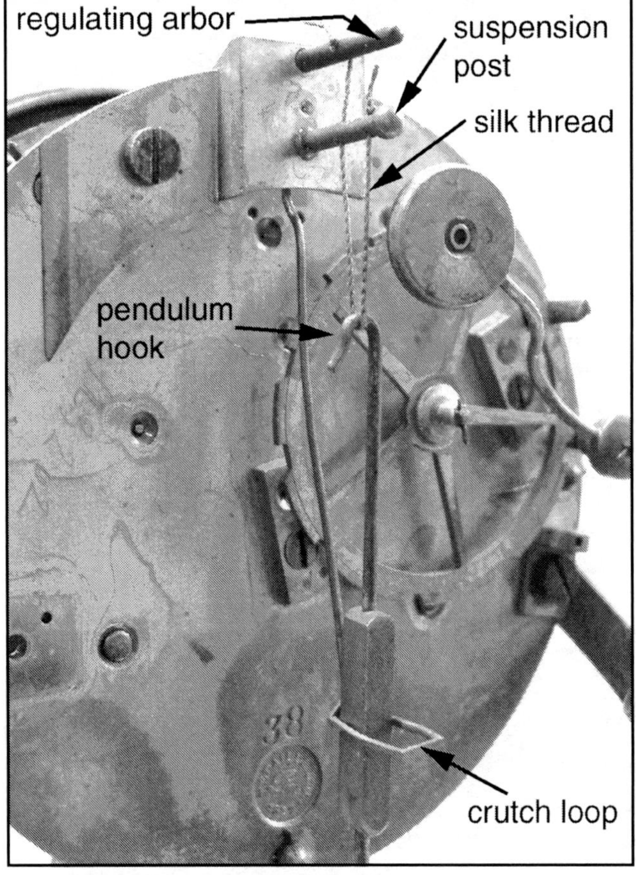

The silk thread suspension has relatively few parts, and they are simple in design. The silk thread is knotted at the regulating arbor and at the suspension post. A needle threader may help in fitting a new thread to a clock. This particular thread has probably been shortened too much through breakage: the clock keeps good time, yet there isn't any thread wrapped around the regulating arbor. It would seem that turning the arbor in either direction would wrap the thread around the arbor, shorten the loop, and make the clock run faster. The extra thread should be longer, so that the thread winds and unwinds as the arbor is turned. Turning one way will speed up the rate, and the other will slow it down. At this writing, at least one parts supplier carries the replacement silk thread. I have been told that it is also possible to locate "nylon bead cord" in a craft store to fulfill the requirement for a thin, pliable cord.

Note that the pendulum hook faces to the left in the photo. It could be arranged to the right just as well, but that would affect the beat of the clock. This poses a possible problem for a customer who takes home a repaired clock with the pendulum detached. If he or she hangs it on with the hook facing the opposite way the repairer had it, the clock may be out of beat and not run well. For this reason, some repairers keep the light pendulum of a silk thread clock attached for transporting it, packing tissue paper around the pendulum to keep it from moving.

The brass block must fit snugly through the crutch loop, but without binding. Lightly oil the block. Ideally, the clock should be keeping good time with the block riding about halfway through the crutch loop, and there should be a few wraps of thread around the regulating arbor.

QUESTIONS AND ANSWERS ON FRENCH CLOCKS

Q. What is the run time for French clocks?

A. Most spring driven examples are 8-day clocks. For best timekeeping, they should be fully wound each week at the same time of day. Some French clocks run for two weeks or nearly so, but I have found that these may lose time the last few days. It is best to recommend weekly winding to your customers.

Q. My rack & snail French clock often stops at 12:14. The hands seem to be stuck. I can advance them with a lot of pressure, until about 6:00, when it gets easier. At this point, the clock will run again. It may run through one more day or perhaps two or three days, before it stops again, always at 12:14. What is the cause?

A. Your problem is that the strike stalls frequently. At 12 o'clock, the rack tail has become jammed against the snail. The actual strike stall-out may occur hours before 12 o'clock, but the jamming happens as the rack tail drags on the back of the snail. Do not continue to force the hands this way. Find out why the strike train stalls, and correct the problem.

Q. I recently assembled a French count wheel movement. I thought the clock was working correctly, until I tested the striking sequence. All was well on the hours and half hours until 8 o'clock. The strike train did not lock after the hour count. It added one more strike. The same thing happened at 8:30; it struck 9 and 9:30 immediately afterwards. I have observed the count lever, and it appears there is not enough room for the lever to drop after the 8 and 9 o'clock hours, thereby causing a skip to the next half hour count, before locking. Help!

A. Try the three other possible positions of the count wheel on its square arbor. Finding the original one will probably cure the problem. Taking apart an old mechanism changes things! It would be a good idea to mark the position of the count wheel on its arbor on your next French count wheel clock. Do not bend the count lever.

Q. I have a French crystal regulator with rack & snail strike. It strikes 8 at 9, and 9 at 10. All the other hours are correct. The snail has a smooth, curved shape, rather than steps. I did notice that it has a bit of a flat on it, but I do not see evidence of any filing. What could be the answer?

A. There is a small variation in the position of the rack tail on the snail that will still allow the strike to count properly. Anyway, a flat area might only cause an extra strike, and you are one stroke short on two of the hours. Try meshing the hour and minute wheels one tooth either way from the present position. Operate the strike and see if there is any change. Be sure to check that 12 and 1 o'clock still work correctly. Look for the punch marks on the motion work wheels. If you line these up, the strike should work correctly. As a last resort, consider moving the rack tail with respect to the rack. This will affect all the hours, but they may still count correctly, and it may correct the 9 and the 10. I would not recommend filing the snail. Filing is a one-way street that will not bring good results.

Q. The strike occasionally stalls, but I cannot find the reason. It will always start up again if I turn the hands around or touch the wheels.

A. It could be any number of things, but check the fly assembly for poise (balance). You see, the whirling fly blade is designed to slip on the arbor as it comes to a stop at the end of each strike sequence. An unbalanced fly stops most times in the same position, and the gear train may not be able to start the blade moving at the next hour or half hour. To check for this condition, let down the mainsprings and separate the plates to allow the fly assembly to come out safely. Set it up carefully with the pivots on knife edges and see if the blade is heavy on one side, stopping with the same side oriented downward each time it is rotated by hand. Run the edge of the heavy side of the blade along a file. After a few strokes, recheck the balance. Do not remove too much brass or do any drastic cutting. Do not apply solder to the lighter side of the blade.

GLOSSARY
OF TERMS

BACK COCK. The supporting plate for the pallet arbor rear pivot, held to the back plate with one screw and a steady pin. This part also supports the SUSPENSION UNIT.

BARREL. Called a going barrel, this is the power unit for a spring driven French pendulum clock. The barrel contains the mainspring, which is hooked to the winding arbor and the outer wall of the barrel. The main wheel is attached to one end of the barrel. The barrel has a brass cover which is not shown here.

BEAT. The ticking of the escapement. To set the beat is to cause its rhythm to be even. The clock's rate is measured in beats per hour.

BELL STRIKE. The most common striking method for French pendulum clocks.

BROCOT ESCAPEMENT. An escapement invented by Achille Brocot. Also called the VISIBLE ESCAPEMENT when placed in front of the dial. The pallets are "D" shaped pieces of polished steel or jewel.

BROCOT SUSPENSION. The most common type of pendulum suspension in the round French movements. A shaft through the dial turns gearing which raises or lowers a brass block that grips the suspension spring, altering its effective length. This makes the clock run faster or slower and it is a fine adjustment. Usually the pendulum bob can be raised or lowered for a gross adjustment.

CLICK AND CLICKWHEEL. The click is a pawl; the clickwheel is a ratchet wheel. Together they allow winding of the mainspring. As the click catches in each tooth of the clickwheel, the winding turns are retained.

CLICKSPRING. A specially formed blue steel spring that keeps tension on the click, seating it in the clickwheel.

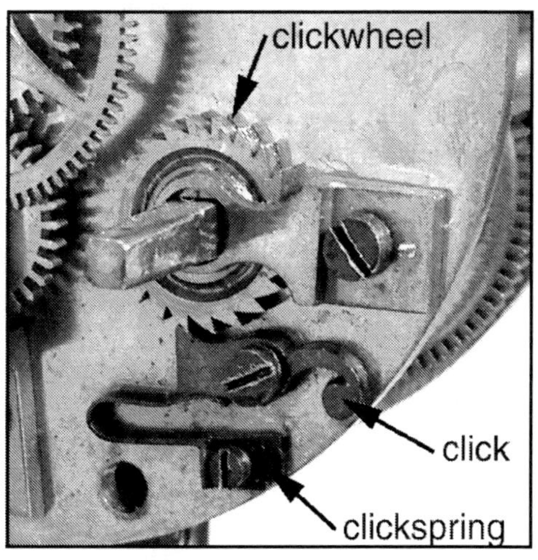

COUNT WHEEL. Also called a locking plate, the count wheel is a device which has the hours 1-12 marked off for the strike mechanism. It counts the hours and half hours, and it cannot readjust itself if the sequence is upset for any reason. The count wheel continued in use on new French movements well into the 19th century but became less common later on. The RACK & SNAIL mechanism, which superceded it, does not require manual correction if the count is upset.

GATHERING PALLET. A small steel piece in a RACK & SNAIL striking clock which moves the rack ahead, tooth by tooth, to count the hours.

GONG STRIKE. A strike mechanism in which a hammer strikes the hours or hours and half hours on a coiled piece of wire held fast in a heavy base. This type of strike is common in French pendulum clocks, but it is not as common as the BELL STRIKE.

HAMMER TAIL. A small projection on the hammer arbor which is raised by the pins on the pinwheel. Very often, a French movement will be found improperly assembled, so that the hammer tail, and hence the hammer itself, is left in the raised position at the end of the strike. It should be at rest.

JAPY FRERES. A large manufacturer of French clocks and movements. Production continued under the Japy name throughout the 19th century and into the 20th, finally declining and ending with World War I. A Japy trademark is a common one seen on round French pendulum movements.

LIGNE. A unit of measurement which is 1/12 POUCE, or 2.256 millimeters. The back plate of most round French movements is marked with the pendulum length stated in pouce and ligne.

LOCKING PLATE. Another name for a COUNT WHEEL.

MARTI, or S. Marti et Cie. produced large numbers of round French movements in the second half of the 19th century and in the early years of the 20th century.

MOTION WORK. A train of gears mounted on the front of a French movement. The center arbor rotates once per hour; the motion work changes this at a ratio of 12:1 to produce the once in 12 hour rotation of the hour hand. In RACK & SNAIL movements, the snail is mounted in this train, on the hour wheel.

MORBIER. A style of clock with a posted movement. The rigid iron framework is made up of top and bottom plates and vertical posts. Steel plates are inserted which carry the pivot holes for the gear trains. These mechanisms are also called Morez or Comtoise. Most were weight driven clocks, but some, such as the one featured in this book, were made as spring driven wall clock movements.

PALLET UNIT. An assembly made up of the pallet body with its two pallets, the pallet arbor, the crutch collet, and the crutch. It is sometimes called by the American name VERGE.

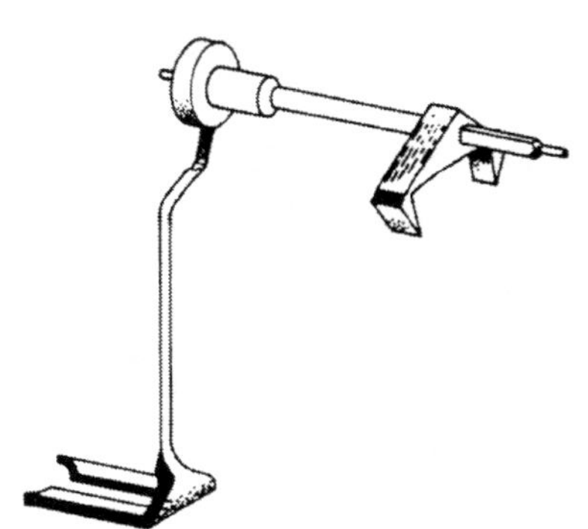

PENDULE DE PARIS. The round French movements produced throughout the 19th century and into the 20th. Many were partially made in outlying districts and finished in Paris.

PICTURE FRAME CLOCK. A style of large French wall clock that originated in the 1870's, from the same area where the MORBIER clocks were made. The movements were of two general types: a square plated striking movement and a posted Morbier movement.

PIVOT. The bearing surface at each end of a clock arbor. Pivot condition, and the wearing of the corresponding bearing hole in the clock plate, are major concerns of the clock repairer.

POUCE. A unit of measurement which is 27 millimeters. Sometimes called a French inch, it was used along with LIGNE (1/12 pouce) to state the pendulum length on round French movements.

RACK & SNAIL. A strike mechanism which has the advantage of being able to correctly count each of the hours, even if the clock has been disturbed by turning the hands or allowing the strike power source to run down. A GATHERING PALLET moves a toothed rack, with the total number of strikes determined by the position of a SNAIL.

REGULATOR: see SUSPENSION UNIT

SILK THREAD SUSPENSION. An early device used in French clocks to support the pendulum with a thin cord.

SNAIL. In a RACK & SNAIL striking movement, the part which determines how many strikes are counted out on the bell or gong. Most French snails are made with 12 distinct steps to represent the hours, but some snails are formed with a continuous curve, as shown here.

SUSPENSION UNIT. A device which supports the pendulum, often including a REGULATOR which permits fine adjustments to the clock's rate. See BROCOT SUSPENSION and VALLET SUSPENSION.

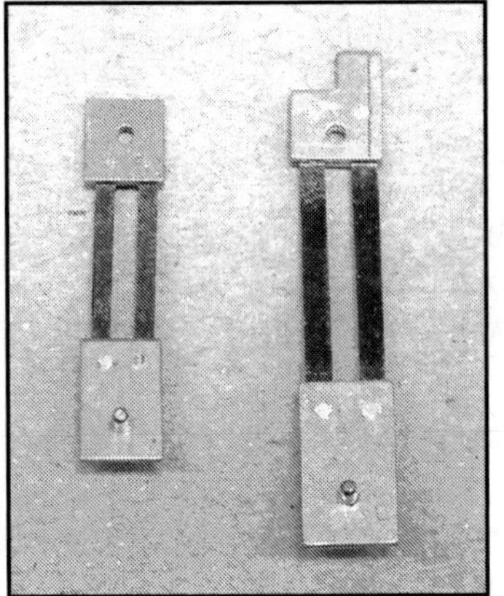

SUSPENSION SPRING. The hanger for the pendulum. It is made up of two blocks of steel or modern plastic, usually with a double spring between them. The one on the left fits many suspension units, including the VALLET SUS-PENSION. The one on the right fits a BROCOT SUSPENSION. The notched piece at the top comes to rest against the adjusting wheel if the pendulum is swung too far back. This prevents the suspension spring from coming free of the suspension unit.

TRADEMARK. A stamped maker's mark sometimes found near the bottom edge of the back plate in a round French movement. The most common trademarks are from Japy Freres, Marti, and Vincenti. A trademark may commemorate an exhibition award, which if marked with a date will at least show that the clock was not made earlier.

WARNING. A short run of the strike train wheels which makes the gear train ready to be lightly "tripped" to begin striking at the hour or half hour. The Morbier movements generally do not use a warning, but most other French pendulum striking clocks do.

VERGE. A term used mostly by American clockmakers to indicate the PALLET UNIT in a clock. In European clocks, the term more properly refers to the PALLET UNIT in a verge escapement, that is, an antique crown wheel and verge.

VALLET SUSPENSION. A variation on the BROCOT SUSPENSION. It has a simpler, more enclosed appearance. It supports the pendulum and allows fine changes in the pendulum length, by means of an arbor passing through the dial.

VISIBLE ESCAPEMENT. See BROCOT ESCAPEMENT.

INDEX